The Political Economy of Agricultural Trade and Policy

The Political Economy of Agricultural Trade and Policy

Toward a New Order for Europe and North America

EDITED BY
Hans J. Michelmann, Jack C. Stabler,
and Gary G. Storey

Westview Press
BOULDER • SAN FRANCISCO • OXFORD

Table 8.1 is reprinted by permission of Oxford University Press from Paul de Hevesy, *World Wheat Planning and Economic Planning in General* (London: Oxford University Press, 1940), copyright © 1940 by Oxford University Press.

This Westview softcover edition is printed on acid-free paper and bound in library-quality, coated covers that carry the highest rating of the National Association of State Textbook Administrators, in consultation with the Association of American Publishers and the Book Manufacturers' Institute.

All rights reserved. No part of this publication may be reproduced or transmitted in any form or by any means, electronic or mechanical, including photocopy, recording, or any information storage and retrieval system, without permission in writing from the publisher.

Copyright © 1990 by Westview Press, Inc.

Published in 1990 in the United States of America by Westview Press, Inc., 5500 Central Avenue, Boulder, Colorado 80301, and in the United Kingdom by Westview Press, 36 Lonsdale Road, Summertown, Oxford OX2 7EW

A CIP catalog record for this book is available from the Library of Congress.
ISBN 0-8133-7992-X

Printed and bound in the United States of America

The paper used in this publication meets the requirements of the American National Standard for Permanence of Paper for Printed Library Materials Z39.48-1984.

10   9   8   7   6   5   4   3   2   1

Contents

List of Tables vii
List of Figures xi
Preface xiii
About the Authors xvii

Introduction 1

Section One: The Political Economy of Agriculture

1   The Political Economy of Agriculture in the European
    Community, *Michael Tracy* 9

2   The Political Economy of Agriculture in Canada,
    Grace Skogstad 35

3   The Political Economy of Agriculture in the United States,
    Gordon C. Rausser 57

**Section Two: The International Agricultural and
Trading Environment**

4   Structural Change in Canadian, United States, and European
    Agriculture, *George L. Brinkman* 95

5   The Crisis in European and North American Agriculture,
    Michele M. Veeman and Terrence S. Veeman 129

6   The GATT: Its Historical Role and Importance to
    Agricultural Policy and Trade, *Tim Josling* 155

Section Three: Prospects for a New World Agricultural Order

7 Prospects for the Uruguay Round in Agriculture, *C. Ford Runge* — 175

8 A New World Agricultural Order? *Murray Fulton and Gary G. Storey* — 195

Section Four: Conclusion

9 Concluding Remarks, *Hans J. Michelmann and Jack C. Stabler* — 233

Tables

1.1	Agriculture in the Economy	16
1.2	Farm Structures	17
1.3	Farm Net Value Added (FNVA) per Agricultural Work Unit (AWU), 1987/88	18
1.4	Farmers' Other Employment (1987)	19
1.5	Self-sufficiency in Foodstuffs	20
1.6	EC Intervention Stocks	20
1.7	Trade in Agricultural Products in the European Community	21
1.8	Community Imports from Third Countries of Products Intended Primarily for Animal Feed	21
1.9	Community Expenditure on Agriculture	24
1.10	"Institutional" Prices—Selected Products	26
1.11	Indices of Producer Prices and Costs	27
2.1	Provincial Sources of Farm Cash Receipts by Commodity, 1987	41
3.1	Annual Gains and Losses from Income-Support Programs Under the 1985 Food Security Act and Trade Restrictions	66
3.2	Productive Versus Predatory Policy Interventions in U.S. Agriculture, 1982-1986 Average	74
4.1	Number of Farms and Average Farm Size, Canada, U.S., and the Euro 10 Economic Community, Selected Years, 1960-86	100
4.2	Farm Numbers and Percentage Distribution by Size in Acres, Canada, Selected Years, 1961-86	101
4.3	Farm Numbers by Size in Acres, United States, Selected Years, 1959-87	101
4.4	Farm Numbers by Size in Hectares of Agricultural Area Used (AA), Euro 10 Economic Community, Selected Years, 1960-85	102
4.5	Percentage Distribution of Farms and Gross Sales by Economic Class, Canada, 1976, 1981, and 1986	103

4.6	Percentage Distribution of Farms and Gross Sales by Economic Class, U.S., 1970, 1980, and 1988	104
4.7	Percentage Distribution of Farms and Gross Margin by Economic Gross Margin Class, Euro 10 Economic Community, 1975, 1980, and 1985	104
4.8	Distribution of Farms and of Farm Output (Gross Sales) by Economic Class, Canada, 1971, 1976, 1981, and 1986	105
4.9	Distribution of Farms and of Farm Output (Gross Sales) by Economic Class, Canada, 1971, 1976, 1981, and 1986 (Inflation Adjusted)	106
4.10	Average Capital Value per Farm, Canada and the United States, Selected Years ($1,000)	108
4.11	Land and Building Values per Hectare, Various Countries 1971-1988	109
4.12	Percentage Distribution of Farms and Land Area Classified by Tenure of Operator and Land Arrangement, Canada, U.S., and the Economic Community, Selected Years, 1969-1987	111
4.13	Trends in the Organizational Structure of Agriculture, Canada and the United States, Selected Years, 1971-1987	113
4.14	Percentage Distribution of Farms by Days of Off-farm Work, Canada and the United States, Selected Years, 1974-1987	114
4.15	Distribution of Employment in Agriculture, Euro 10 Economic Community, 1975, 1980, and 1985	115
4.16	Selected Farm Data Classified by Indexed Sales, Canada, Selected Years	117
4.17	Indices of Prices Paid and Received by Farmers (1975 = 100), Canada, U.S., and the European Community, Selected Years, 1975-1988	122
4.18	Trends in Self-sufficiency Percentages, Selected Countries, 1970-1988	124
5.1	Key Indicators for Selected Developed Nations and Their Agricultural Sectors	130
5.2	Economic Trends in Western European Agriculture, 1977 to 1989	132
5.3	Economic Trends Relating to Agriculture, United States, 1973 to 1989	134
5.4	Economic Trends in Canadian Agriculture, 1971 to 1990	136

5.5	Producer and Consumer Subsidy Equivalents by Nation, 1984 to 1988	137
5.6	Total Transfers Associated with Agricultural Policies in Developed Nations	138
5.7	Direct Payments (Government Subsidies) to Agriculture, Canada and the Prairie Provinces, 1971, 1975, and 1978 to 1990	139
8.1	Seeded Acreage of Wheat—United States, Australia, Canada and Argentina	202
8.2	Wheat Production, Exports and Market Share, U.S. and EC, 1960-1988	213
8.3	Historical Wheat Prices, U.S. and EC, 1960-1989	217

Figures

3.1	Average Direct Government Payments per Farm by Sales Class, 1988	71
3.2	World Trade in Wheat, Coarse Grains, and Soybeans, 1972-1990 Crop Years	82
4.1	Distribution of Farms and Sales of Agricultural Commodities by Agricultural Sales Class, 1974, 1978, and 1982	107
4.2	Index of Relative Change in Land Values (1973 = 100)	110
4.3	Index of Multifactor Productivity Change, Canada and the U.S. (1977 = 100)	121
4.4	Farming: Myth and Reality	125
5.1	The Range of Policy Choices in European and North American Agriculture	148
5.2	Some Prevalent Myths that Relate to Agriculture in North America and Western Europe	152
7.1	Export-Distorting Policies	177
7.2	Import-Distorting Policies	178
7.3	Output-Distorting Policies	179
7.4	Oilseed Planting Trends	184
7.5	U.S. Soybean Exports and U.S. Soybean Meal Exports	185
8.1	Chronology of European–North American Agricultural Policies, 1800-1990	198
8.2	Two-Region Policy Transition Model, 1850-1990	206
8.3	The U.S. and EC Wheat Market in the 1960s	210
8.4	The U.S. and EC Wheat Market: Mid-1970s versus the 1960s	212
8.5	The U.S. and EC Wheat Market: Late 1970s/Early 1980s versus the Mid-1970s	214
8.6	Economics of Protection in the EC	216

8.7	The U.S. and EC Wheat Market: Mid-1980s versus the Late 1970s/Early 1980s	219
8.8	The U.S. and EC Wheat Market: Late 1980s versus the Mid-1980s	220
8.9	The U.S. and EC Wheat Market: Free Trade versus the Late 1980s	221
8.10	The U.S. and EC Wheat Market: Free Trade versus the Late 1980s, Inelastic Supply in the EC	223
8.11	The U.S. and EC Wheat Market: Production Control versus the Late 1980s	224

Preface

The chapters of this volume are the revised versions of papers presented at a conference entitled "The Political Economy of European–North American Agricultural Policy and Trade," held March 7-9, 1990, in Saskatoon, Saskatchewan, Canada. The conference was one of a series held annually by the Canadian Council for European Affairs on issues of importance for Canadian–European relations, often with the cooperation of institutions having compatible objectives. Hence the co-sponsorship of the conference by the University of Saskatchewan's Department of Agricultural Economics and Department of Political Studies.

The subject matter of the conference and volume is particularly timely. For most of the 1980s and at least until the conclusion in the near future of the Uruguay round of GATT negotiations, if not well beyond that time, one of the most perplexing issues for economic relations among the European Community, the United States, and Canada, was, and will continue to be, the rationalization of international agricultural trade. That issue is at the very heart of the present round of GATT negotiations. If it is not resolved, it may well impede the resolution of other trade issues and thus doom the Uruguay round to failure because of the connection made by many participants between solving agricultural trade problems and solving other trade issues.

Saskatoon was chosen as the venue of the conference not only because its organizers are affiliated with the University of Saskatchewan but also, and more importantly, because agriculture is central to the economy of Saskatchewan and because much of the province's agricultural production is destined for export. Hence the organizers, who are also the editors of this volume, felt that the Saskatoon context and ambience would be appropriate for the meeting and that its subject matter would be attractive to a large number of prairie agricultural producers and producer organizations. It is because the conference was held in Saskatoon that the concluding remarks, particularly those by Jack C. Stabler, are directed toward the implications of the presentations for Canada and Saskatchewan.

It is clear, however, that the subject matter of the conference and this volume are also relevant to a much wider audience, viz., Europeans and Americans and indeed all those interested in agricultural policy and trade worldwide. Areas outside the North Atlantic triangle were not explicitly examined at the conference because of its focus on European–North American relations and because of time and resource constraints. Nonetheless, the magnitude of agricultural exports by the EC, the U.S., and Canada and the importance of the interactions among these three for international agricultural trade makes this volume significant for anyone interested in trade issues.

The production of an edited volume based on papers presented at a conference involves a large number of persons and organizations and, for the editors of the volume, engenders an equally large number of debts, which we hereby willingly acknowledge. We are grateful, first of all, to all those in the University of Saskatchewan and the Canadian Council for European Affairs, the two primary sponsoring organizations, who helped us in planning, organizing, and holding the conference. Foremost among those at the university are Murray Fulton of the Department of Agricultural Economics and George James of the Division of Extension and Community Relations. The division provided invaluable infrastructural help, without which a major conference cannot be efficiently organized and held. We also wish to thank the university's colleges of Agriculture and of Arts and Science for financial support and encouragement.

Individual members and donor organizations of the Canadian Council for European Affairs who deserve special recognition for their support and advice include Ambassador John G.H. Halstead and Professor Panayotis Soldatos, council chairman and vice-chairman, respectively. Among the primary conference donor organizations was the Delegation of the Commission of the European Communities in Ottawa, whose head, Ambassador Jacques Lecomte, encouraged the whole project from its initial phases, helped to assure a secure financial base at the outset, and then participated well beyond the call of duty in Saskatoon. Roy Christensen, head of Press and Information at the delegation, contributed to the project during all its stages. We are grateful to External Affairs and International Trade Canada, a second major donor, and to Agriculture Canada, which also provided substantial funding. The Ministry of International Affairs, Government of Quebec, the Ministry of Intergovernmental Affairs, Government of Ontario, and the Bank of Montreal contributed as donor organizations of the council.

We particularly wish to thank our co-authors, because without their contributions neither the conference nor the present volume would have been possible. We are grateful, also, for the expeditious manner in which manuscripts were delivered to us for editing. Nora Russell expertly read all manuscripts, making improvements in clarity and assuring consistency in style. We are grateful to Mary Frances Schmidt for typing a camera-ready copy of the manuscript. Lyle Yuzdepski aided in proofreading and walked many miles in transferring chapter manuscripts between the offices of various participants in the editorial and production processes. Finally, we wish to thank Marykay Scott of Westview Press for advice and encouragement.

Hans J. Michelmann
Jack C. Stabler
Gary G. Storey

About the Authors

George L. Brinkman received his Ph.D. from Michigan State University. As Professor of Agricultural Economics at the University of Guelph, he specializes in agricultural policy, farm structure and rural development. He has published widely on topics of structural change in Canadian agriculture. He has been President of the Canadian Agricultural Economics and Farm Management Society and Councillor to the Agricultural Institute of Canada.

Murray Fulton received his Ph.D. from the University of California, Berkeley, earlier receiving degrees from the University of Saskatchewan, where he was Rhodes Scholar, and Texas A & M University. He is currently Associate Professor of Agricultural Economics and associated with the Centre for the Study of Cooperatives at the University of Saskatchewan. He lectures on econometrics, cooperatives and industrial organization and has recently co-authored a study for the Economic Council of Canada entitled Canadian Agricultural Policy and Prairie Agriculture.

Tim Josling obtained his Ph.D. from Michigan State University, with earlier degrees from the University of London and Guelph University. He has taught at the London School of Economics, Reading, and is currently Professor at the Food Research Institute, Stanford University. His specializations are in trade and agricultural policy. In 1985 he co-authored a book on agricultural policies and world markets. A forthcoming book entitled *Agricultural Policy Reform* is being co-authored with Wayne Moyer.

Hans J. Michelmann received his Ph.D. from Indiana University. He is Professor of Political Studies at the University of Saskatchewan, Director General of the Canadian Council for European Affairs, and Editor of the *Journal of European Integration*. He specializes in European and comparative politics, including the politics of the European Community, Canadian politics and European, especially German, politics.

Gordon C. Rausser received his Ph.D. at the University of California, Davis. Following appointments at Harvard and Iowa State he is currently the Robert Gordon Sproul Distinguished Professor of Agricultural and Resource Economics, University of California, Berkeley. His areas of specialization include public policy in food, agriculture and natural resource systems. He and his students have won several professional awards. He has served as Department Chairman and Editor of the *American Journal of Agricultural Economics*. He has served on several national commissions and the President's Council of Economic Advisors (1986/87) and as Chief Economist at the Agency for International Development since 1988. Professor Rausser has published over 200 articles, monographs and books.

C. Ford Runge received his Ph.D. from the University of Wisconsin. He is Associate Professor in Agricultural and Applied Economics at the University of Minnesota. He specializes in national farm and economic policy and land economics. In 1976-77 he worked on the American House Agricultural Committee staff and in 1988 he served as a Special Assistant in Geneva to the U.S. Ambassador to the GATT. "Agricultural Trade in the Uruguay Round: Into Final Battle" is a recent publication. He is currently Director of the Centre for International Food and Agricultural Policy.

Grace Skogstad is an Associate Professor of Political Science at the University of Toronto. She received her Ph.D. degree at the University of British Columbia in 1976 and has done extensive research on agriculture and agricultural trade. Among her recent publications are: *The Politics of Agricultural Policy Making in Canada*, University of Toronto Press, 1987; *Agricultural Trade Policy: Domestic Politics and International Trade*, IRPP, 1989 (co-edited with Andrew Fenton Cooper); "The Application of Canadian and U.S. Trade Remedy Laws: Irreconcilable Expectations," *Canadian Public Administration*, Winter, 1988.

Jack C. Stabler obtained his Ph.D. from the University of Utah. He is currently Professor and Head of the Department of Agricultural Economics at the University of Saskatchewan. His specialization is in regional and rural development. Recent publications focus on benefit-cost methodology, as used to evaluate water resource development projects, and on the role of service industries in the development of regional and rural economies.

Gary G. Storey received his Ph.D. from the University of Wisconsin. He is currently Professor of Agricultural Economics at the University of

Saskatchewan, where he specializes in agricultural marketing, trade and policy. Recent publications focus on commodity analysis in grains and oilseeds, including a co-authored study for the Economic Council of Canada on grain market outlook. He has been President of the Canadian Agricultural Economics and Farm Management Society and Councillor to the Agricultural Institute of Canada.

Michael Tracy was educated at Cambridge University where he received a B.A. in modern languages and an M.A. in economics. His professional career has been largely in government service in the United Kingdom and OECD; from 1973 to 1983 he was Director in the Secretariat of the Council of Ministers of the European Communities in Brussels. He has had a parallel career with numerous academic appointments. He is currently visiting professor at the Karl Marx University in Budapest. He has published widely on European agricultural history and policy. *Government and Agriculture in Western Europe 1880-1988,* (New York State University Press, 1989) is his most recent publication.

Michele M. Veeman received her Ph.D. from the University of California, Berkeley, with earlier training from Massey University, New Zealand, and the University of Adelaide, Australia. She is Professor of Agricultural Economics in Rural Economy, University of Alberta. She lectures and has written widely on agricultural marketing and policy topics. She is currently President-Elect of the Canadian Agricultural Economics and Farm Management Society.

Terrence S. Veeman, educated at the University of Saskatchewan where he was Rhodes Scholar, received his Ph.D. from the University of California, Berkeley. He is currently Professor in the Department of Economics and Rural Economy, University of Alberta. His research areas include agricultural policy, trade, and land and water use issues. He has co-authored, with Michele Veeman, "The Future of Grain: Canada's Prospects for Grains, Oilseeds and Related Industries."

Introduction

The 1980s have been turbulent years for the agricultural sectors of most economies. Global agriculture has undergone substantial change and there are now concerns over the environment, increased international competition, declining prices and farm incomes, and increased pressure for further government intervention. The period 1985 to 1990 has been marked by the escalating trade war between the European Community (EC) and the United States (U.S.). The major cause of this impasse is the grain surplus that has resulted from the high price-support structure of agricultural policy in the EC and other exporting countries.

The situation is comparable to the disruptions in agricultural trade that took place prior to the 1930s, which resulted from grain surpluses in the four major exporting countries—Canada, the United States, Australia, and Argentina. These events, which led to disastrous conditions for farmers worldwide, were instrumental in causing governments to reassess their positions toward agriculture. The 1930s were a transition period during which governments began to intervene in the agriculture sectors of most developed economies, resulting in the abandonment of free trade. In the aftermath of the Second World War, agriculture and food came to be treated differently from other goods in terms of international trade policy. At the insistence of the U.S., for example, agriculture was excluded from the GATT when it was formed in 1947.

The year 1990 could prove to be pivotal for the future agricultural policies of Europe and North America. The GATT negotiations of the Uruguay Round will conclude in December. For the first time since GATT was formed, agriculture has been included as a main negotiation issue. The focus is on government support and protection policies, in the form of subsidies, that have been extended to agriculture.

Policy makers have come to realize, particularly given the crises in agriculture of the 1980s, that government intervention is not the solution to agriculture's problems. Instead of raising farm incomes, government subsidies have tended to be capitalized into higher land values, resulting in increased farm debt and lower farm incomes when the debt proved difficult to service in periods of declining prices. In addition, higher price supports and payments have encouraged greater technological development and thus increased productivity, which has contributed to food surpluses. At the same time, the lack of economic development in the third world has limited the effective demand for food, which could have reduced these surpluses in the developed countries.

Policy makers, therefore, have begun to look at the possibilities for the return to a free trade environment coupled with less overall government intervention. This is essentially the position taken by the U.S. and to a similar extent, the Cairns Group, at the GATT negotiations. But there is no unanimity among the negotiators that this is the right solution. The EC in particular has resisted a movement in this direction. What, then, is the correct solution and "order" for world agriculture?

Agricultural policy has always resulted from a complex interaction between economic and political factors. The expanded level of trade in the 1970s, the shift from fixed currency to floating exchange rates, increased interest rate instability, and the closer linkages of world commodity markets, have added to the complexity. As a result, domestic agricultural policy cannot be developed in isolation from that of other countries, and therefore a more comprehensive approach to policy analysis is needed.

The current GATT negotiations have prompted studies on the effects of trade liberalization. Most of these have been economic in nature. One of the difficulties in studying agricultural policy, however, is that a well-informed analysis requires both an economic and political perspective; most studies have tended to focus on only one of the two. It can be argued, for example, that the failure to adequately understand the influence of the farm lobby and farm political power has limited economists' ability to analyze policy and predict policy developments. This applies in particular to the EC, where the power of the farm bloc in both France and Germany has had a strong influence on shaping agricultural policy. Evidence from studies and conferences, including G7-level summits, indicates that the political realities are not well understood. This realization provided the basis for the conference held in Saskatoon, Saskatchewan, in March 1990.

The current trend is toward greater overall world trade and hence specialization. World trade is equal to approximately one-fifth of gross world product. Growth and prosperity are linked to trade, and it is for this reason that policies that distort efficient agricultural production and trade must be reviewed and, if necessary, replaced by more effective measures. The purpose of the conference and hence the objectives of this book are: (1) to examine agricultural and trade policy in the EC, the U.S., and Canada from a political economy perspective; (2) to examine the structural changes that have taken place in the EC, the U.S., and Canada, and the resulting agricultural crisis; (3) to examine the impact of GATT on agricultural policy and to explore the context and process of the current GATT negotiations for trade liberalization; and (4) to examine the prospects for a new "order" for agriculture.

Structure of the Book

The book contains nine chapters organized into four sections. Section One presents a descriptive analysis of the political economy of the EC, the U.S., and Canada. Section Two describes the structural changes and the crises in agriculture. Section Three explores the prospects for a new world agricultural order, first, by focusing on the potential outcomes of the Uruguay Round of GATT, and second, by examining the prospects for a new "order" for agriculture. Section Four presents a set of concluding remarks that first highlights the impact of political factors on agricultural trade and policy as outlined in the preceding chapters, and then discusses the implications of the findings for Saskatchewan.

Section One: The Political Economy of Agriculture

The three chapters in this section trace the development of agricultural policy in each of the three settings, with particular focus on the political and economic forces that have given rise to current agricultural and trade policy positions. The chapters also explore the forces that will determine future policy choices, thus influencing the structure and characteristics of the domestic agricultural sector and trade. In Chapter 1, Michael Tracy, examining the political economy of agriculture in the European Community, stresses that the prime motive for the formation of the EC was political. He focuses on the political aspects of the formation of the Common Agricultural Policy (CAP) and examines the wider economic issues. He also discusses CAP reform. In Chapter 2, Grace Skogstad examines the agricultural policy process in Canada, and the relative importance and power of the federal and provincial governments and producer groups in shaping agricultural policy. Following a discussion of national and international economic and political realities, she concludes that change leading to the national government assuming greater autonomy relative to producers in Canadian agricultural policy is unavoidable. In Chapter 3, Gordon Rausser presents an analysis of U.S. agricultural policy where the public sector involvement in agriculture is seen to have been both "productive" and "predatory." He argues that U.S. agriculture is more complex than the extremes posed by welfare economics and rent seeking. He sees an integration of the—predatory and productive—type policies that provides a basis for understanding U.S. agricultural policy.

Section Two: The International Agricultural and Trading Environment

The three chapters in this section examine the implications of the political and economic factors presented in Section One, offering a descriptive analysis of structural changes and crises in the agricultural economies of the EC, the U.S., and Canada. The section focuses particularly on GATT as an institution and on its impact on agricultural policy and trade in the post–Second World War era.

In Chapter 4, George Brinkman provides a description and assessment of structural changes to agriculture in Canada, the U.S., and the EC. Included in the chapter are an examination of structural policies and major structural dimensions and conditions at the farm level, a survey of macrostructural characteristics, and finally, implications for policy. In Chapter 5, Michele and Terry Veeman focus on the roots of the crises in the agricultural economies of Europe and North America. They detect similar patterns of declining farm numbers and farm incomes in both settings and see farm family incomes falling behind national incomes. They outline the pressures for reform and examine policies that would seem appropriate to solve the problems, concluding that the agricultural sectors of developed countries must become more market-oriented and market-driven. In Chapter 6, Tim Josling reviews the GATT and examines its role and importance to agricultural policy and trade. He provides a detailed discussion of the events leading to the Uruguay Round, and examines the context in which the GATT could arrive at an agreement that might provide a "turning point in international cooperation."

Section Three: Prospects for a New World Agricultural Order

Section Three takes a more specific look at the GATT negotiations and the implications of its success or failure for future agricultural policy. The prospects for a new agricultural order are examined from a historical perspective and from the perspective of a modelling framework.

In Chapter 7, C. Ford Runge examines the prospects for success or failure of the GATT negotiations over agriculture, specifically for a "core" agreement that might emerge. He discusses the interaction of the U.S. 1990

Introduction 5

Farm Bill with the GATT negotiations, and the ways in which the Uruguay Round could affect subsequent trade negotiations. He concludes that there is likely to be an eventual compromise that will move agricultural economies toward trade reform. He warns, however, that increased threats to liberalized trade reform will arise over environmental and food quality concerns. In Chapter 8, Murray Fulton and Gary Storey take a broader look at agricultural policy reform by first examining the historical development of agricultural policy and trade from 1800 to the present. Second, they examine the prospects for different "orders" for agriculture, including a return to free trade, by developing a model of EC-U.S. agriculture as seen over the past twenty-five years. The conclusion is that a return to completely free trade is unlikely, given current political/economic realities.

Section Four: Concluding Remarks

In this final section, a summary assessment of the conference that formed the basis for the book is presented, along with a discussion of the implications for Saskatchewan, the geographical setting in which the conference was held.

In Chapter 9, Hans J. Michelmann highlights the importance of political factors for agricultural policy and trade. J.C. Stabler then explores the implications for Saskatchewan. This approach is not idiosyncratic; in fact, given Saskatchewan's unique situation amongst the agricultural trading areas of the world, such a discussion highlights the dislocation that the present trading regime entails, and also the potential payoffs involved in reform of the present system. Hence, Saskatchewan is a useful case study, and a discussion of its problems and prospects is an appropriate ending for the book.

SECTION ONE

The Political Economy of Agriculture

1

The Political Economy of Agriculture in the European Community

Michael Tracy

This chapter discusses the development of agricultural policy "with particular focus on the political and economic forces which have given rise to current agricultural and trade policy positions," and explores "the forces which will determine future policy choice."

In the case of the European Community (EC), this task is not easy. In any country, agricultural policy results from a complex interaction between numerous forces: one might say broadly that politicians take decisions in the light of their assessment of the likely reactions of farmers and other interest groups, and subject to a variety of constraints. An examination of the European Community also has to take account of the different interests of twelve member states, in whose economies agriculture plays widely differing roles, and in some of which, moreover, there are wide regional differences.

This chapter deals first with the most important national attitudes and explains how they influenced the formation and development of the Common Agricultural Policy (CAP); it then discusses the recent reforms, and tries to explain the background to the EC proposals in the context of the Uruguay Round.

National Interests and the Evolution of the CAP

The Founding Members

It should never be forgotten that the prime motive for the formation of the EEC was *political*: in less than a century, France and Germany had fought each other three times, twice in the context of European-wide and ultimately world-wide wars. Already in the 1950s, the European Coal and Steel Community had bound together these key industries of the six original member states, and above all those of France and Germany. The European Economic Community was seen as a further vital step along this path.

Agriculture, however, presented a problem. Each of the Six already protected its farmers, but to varying degrees and in varying ways. But agriculture could not be left out of the "common market." French agriculture, in particular, was already producing surpluses for which outlets had to be found; and in the expansion of agricultural exports, France saw a valuable source of foreign exchange. Moreover, France feared German competition in the industrial sector; advantages in agriculture were the necessary quid pro quo, and for France this meant including agriculture in the common market. But the powerful French farm organizations were not prepared to give up the protection and support that French agriculture had enjoyed for many years; so protection against imports from third countries had to be maintained, and common market organizations had to replace national ones.

The German Federal Republic, with a strong political and economic interest in the creation of the EEC, was ultimately prepared to concede the essential French demands. It too, however, had its farmers to worry about; although Germany was highly industrialized, and the agricultural population was relatively small, farmers' political strength was considerable. This was largely because of the close links between the main farm organization, the *Bauernverband*, and the ruling Christian Democratic party (CDU). So for Germany, too, continued support and protection was essential; moreover, the common price level could not be too much below the existing national level.

The Netherlands, like France, had a strong interest in access to the German market for its agricultural exports. As a highly competitive producer, the Netherlands might have preferred a relatively low price level. However, in the Netherlands, too, agricultural markets were quite extensively organized, and the Netherlands was prepared to acquiesce in setting up interventionist common market organizations.

In Belgium and Luxembourg, farmers were accustomed to high levels of support. In both countries, in spite of a high level of industrialization, farmers remained powerful. In Belgium this was—and remains—due to close links between the *Boerenbond* and the Flemish Christian Democratic party (CVP).

For Italy, an agricultural common market offered the prospect of improved outlets for its exports of agricultural produce. But due to political and economic weakness, Italy exerted relatively little influence on the early development of common market organizations.[1]

So the CAP was not created in a vacuum, but was an amalgam of existing national measures. The system of variable levies on imports from third countries, together with intervention boards to maintain a steady price level within the EC, owed most to the German *Einfuhr- und Vorratsstellen.* The system of export "refunds" was thrown in almost for good measure; at the time, with the Community of Six still a net food importer, this was not seen as a major issue. The common price level was fixed, after much argument, somewhat nearer the high German level than the lower Dutch or French ones.

For France, a key element was to ensure that the financial responsibility for market support would lie with the EC; this aim, too, was realized with the first regulations, adopted in January 1962. The French were then able to claim that three essential "principles" had been accepted: the unity of the internal EC market, preference for EC suppliers over third countries, and financial solidarity among the member states for agricultural support. Further, the ultimate success in reconciling such different national positions enabled the CAP to be described as a "pillar" of the European Economic Community. Invocation of the principles and of the pillar concept became part of the rhetoric of the CAP, subsequently to be invoked whenever the CAP was challenged.[2]

For a long time, the French remained quite fiercely attached to the CAP, resisting pressures for change from every quarter. Even the word "reform" was taboo for some considerable time, and without the French, not much could be done about the CAP. In annual price reviews during the 1970s, the French minister of agriculture was usually among those pressing for price increases. This may have been a mistake; French interests might in the long term have been better served by keeping down the price level in order to force out German producers, but almost all the militant French farm organizations continued to demand price increases at every opportunity.

For some time the British were inclined to hold the French responsible for the protectionist aspects of the CAP; "inefficient French peasants" were supposed to be at the root of the problem. But this was only partly true. A perceptive article in *The Economist* of November 5, 1977, made the point: "Critics of the EEC's common agricultural policy normally blame the French for its wilder absurdities. But they are wrong. The real blame lies with the Germans." Reference was made above to the paradoxically strong position of farmers in the Federal Republic, due initially to the link between the *Bauernverband* and the CDU. Later, when Socialists (SPD) and Liberals (FDP) formed a coalition, the very farm-oriented FDP minister of agriculture, Josef Ertl, was usually able to get his own way because his party was essential to the coalition. His successor, Ignaz Kiechle, another Bavarian farmer, was also in a strong position, representing the minority Bavarian wing (CSU) in a CDU-led government. Although by the 1980s the farm population in the Federal Republic was down to about 5 percent of the total, politicians still considered it significant enough to sway the voting outcome in a number of marginal constituencies.[3]

A word must be said at this stage about the "agri-monetary" problem. This is impossible to explain in a few sentences, and is mainly an internal EC issue. However, it is necessary to point out the implications for Germany, as a result of the successive revaluations of the mark. Under CAP rules, common agricultural prices in European currency units (ECU) should then have been translated into reduced prices in marks for German farmers. The so-called "green rates" and "monetary compensatory amounts" (MCAs, and other devices too complicated to mention here) have served to avoid or mitigate this effect, but at the cost of constant friction. At every price review since the early 1970s, the French and others have attacked Germany on this issue, with limited success. The MCAs, moreover, require controls at the EC internal frontiers that are inconsistent with common market principles and in particular with the 1992 single-market aim, a problem that at the time of writing remains unresolved.

The British Problem

The United Kingdom (UK), having opted out of the preparatory talks on the EEC, was in no position to influence the negotiations among the six on the establishment of the CAP; and when the British Conservative government under Ted Heath finally took Britain into the EEC in 1973, it had to accept the CAP as it then stood, with only minor concessions and temporary derogations. This meant abandoning the "deficiency-payments" system, and accepting the import-levy system in place of virtually free food imports, with a consequent rise in consumer prices; it also meant accepting "Community preference" in place of "Commonwealth preference."

This absorption of the large British market into the protective EC market had dramatic implications for agricultural trade—for Commonwealth countries (grain from Canada and Australia, dairy products from New Zealand) and for other world exporters (the U.S., and perhaps, above all, Argentina and Uruguay for beef). The UK did obtain a concession on New Zealand butter, and has contrived to maintain this—though for much reduced quantities—to the present day. It is arguable, however, that the bargaining power periodically exercised in pushing through this concession restricted the UK's ability to gain other negotiating points.

With the growing budgetary cost of the CAP—which will be discussed shortly—the major UK concern became its high net contribution to the EC budget. Because the UK is still a major food importer, it has more import levies to pay into the EC budget than anyone else, while the relatively small size of its farm sector limits the financial benefits it can draw back from the farm fund or FEOGA (much the biggest element in the budget).

The British demand for a special budgetary concession emerged as the main issue in the renegotiation attempted by the Labour Government, which returned to power in 1974; and it has been vigorously pursued by Margaret Thatcher since she became prime minister in 1979. Complex budgetary arrangements have been devised—with great reluctance by the other partners—to alleviate this problem. The relevant point for the present purpose is that this factor made Britain the main opponent of increases in the common prices for agricultural products at the annual reviews. As will be seen below, however, it was not till the mid-1980s that this position met with much success. It may be that other delegations that might have shared British concerns were reluctant to side too openly with the UK, whose commitment to the European cause was still seen as doubtful. Unlike the French, the British have rarely managed to present national demands as being "in the Community interest."

It should be observed that the efficient UK farms were well able to respond to the stimulus of higher prices under the CAP—and were indeed encouraged to do so by the government, as this helped to increase Britain's receipts from the EC farm fund. It is also to be noted that since the pound sterling has tended to fall against the ECU, British farmers have been fortunate to be in the opposite position from their German counterparts, *benefiting* from price increases in sterling terms, although here the "green rate" system has been used to delay these adjustments. This meant that frequently the British minister of agriculture was in the position of pressing for price cuts for everyone else, knowing that his own farmers could get a comfortable price increase!

For the two other countries that joined the EC in 1973—Ireland and Denmark—access to a wider market for their agricultural exports was of vital importance; Ireland had been much too exclusively dependent on the British market, and Danish exports had remained relatively stagnant as compared with those of the Netherlands. Ireland became a fervent supporter of the CAP, always ready to support France in the Council of Ministers. Denmark, too, in spite of its previous noninterventionist principles, found its interests well served by the CAP; its representatives generally concentrated on eliminating the remaining obstacles to intra-EC trade, including the MCA system.

The Southern Dimension

It was mentioned earlier that Italy had not played a very significant role at the formation of the CAP. By the mid-1970s, however, the Italian minister of agriculture, Giuseppe Marcora, was complaining with increasing bitterness that his country was not doing well out of the CAP. His complaints related to both market policy and "socio-structural" policy.

As regards market policy, Italian agricultural exports were not doing as well in the rest of the EC as had been expected. Although this was largely due to poor quality and inefficient marketing, inadequate market support was blamed. Another cause of Italian concern arose from the EC's efforts to develop a "global Mediterranean policy," involving trade concessions to the third countries concerned around the Mediterranean basin, which inevitably included agricultural products in which Italy, too, was interested.

Italian pressure in the Council of Ministers of Agriculture—using the negotiating power provided by the practice of seeking unanimous decisions, discussed below—led to significant strengthening in the arrangements for protection and support for "southern" products, in particular fruit and vegetables, wine, and olive oil. One result was that these as well as the "northern" products—cereals, sugar, milk, and beef—soon became expensive items for the EC farm fund.

The evolution of socio-structural policy is not of direct concern to an overseas audience. Suffice it to say that a common socio-structural policy hardly existed until 1972, and that the measures then adopted proved largely ineffective. They were scarcely applied in Italy, partly through administrative failures, and partly because very few farms were big enough to qualify for investment aids. But in 1978 and 1979, Italy obtained a "Mediterranean package," which gave Italy and southern regions of France a number of special programs to promote structural and infra-structural improvement (Ireland took the opportunity to get some measures of its own, too, and the process proved contagious).

When Greece joined the EC in 1981, similar programs were extended to it. Greece later insisted—as the price for accepting the accession of Spain and Portugal—on the adoption of ambitious "Integrated Mediterranean Programs." In due course, Portugal and Spain also obtained special support for structural improvement. It should be noted that in the Community of Twelve, the southern member states have enough votes to *block* decisions in the Agriculture Council and thus have substantial bargaining power.

The Community of Twelve thus incorporates a north-south element, in which the less prosperous southern members expect financial transfers from the north. This is a situation the Treaty of Rome had not foreseen; it had envisaged a community with partners at approximately equal stages of development, and the EC budget has had to be adapted in consequence, as will be indicated later.

Economic Issues

So far, this chapter has been discussing mainly political forces; it is time to introduce a few facts and figures, to illustrate the problems facing our decision makers.

Farm Structures and Incomes

Though the point may be obvious, it is necessary to reiterate that farm structures in Western Europe are not the result of colonial implantation, but have been formed over many centuries. In consequence they are, in most regions, economically irrational, but they cannot be changed quickly.

In fact, agriculture's share in employment has been much reduced, though in most countries it is still excessive by comparison with agriculture's contribution to the national income (Table 1.1). There have been large reductions in the number of farms (Table 1.2). This is not due as much as we would like to think to our socio-structural policy, but mainly to economic forces pulling young people into better-paid, nonfarm employment, and to demographic forces causing many small farms to disappear as their former owners die or retire. But we still have, in most regions, too many small farms. The averages in Table 1.2 give some idea of the contrasts between our member states; in terms of "European size units" (a more helpful measure than area), the Netherlands, the UK, Denmark, and Belgium come out on top, Ireland and the southern European countries at the bottom. Note the mediocre rating of Germany.

The range within each country is also significant (Table 1.3—here the measure is "farm net value added" per "agricultural work unit"). The category below 4,000 ECU (5,500 Canadian dollars or 4,760 U.S. dollars, at January 1990 exchange rates) is uncomfortably large almost everywhere except in the three Benelux countries. More and more farm families do not

Table 1.1 Agriculture in the Economy

	\multicolumn{4}{c}{Employment}	Gross Domestic Product	
	In Agriculture (thousands)		Agriculture as Percentage of Total		Agriculture as Percentage of Total	
	1958	1988	1958	1988	1958	1987
Ireland	407	166	38	15.4	26	10.3
United Kingdom	1030	581	4	2.2	4	1.7
Denmark	380	169	20	6.3	16	4.0
Netherlands	495	245	13	4.7	11	4.1
Belgium	324	100	9	2.7	7	2.2
Luxembourg	24	6	18	3.4	9	2.4
France	4459	1428	24	6.8	11	3.5
Germany (BRD)	3978	1327	16	5.2	7	1.5
Italy	6974	2058	33	9.9	19	4.5
Greece	1899[a]	971	57	27.0	28	15.6
Spain	4547[a]	1695	42	14.4	25	5.2
Portugal	1381[a]	887	43	20.7	27	6.4

[a] 1960

Sources: OECD, *Agriculture and Food Statistics 1955-1968*; and Commission, *The Agricultural Situation in the Community*, 1989 report (Table 2.0.1).

depend on farm incomes alone (Table 1.4), though we do not have precise overall information on the amount of these other incomes. But even taking this into account, there is still a widespread problem.[4]

Farm income statistics, which are always an average of some kind, are sometimes difficult to interpret. The following is a specific case. The author lives in an area of Belgium—the Brabant Wallon—where the soil is fertile and the farms quite large (usually in the thirty-to-sixty-hectare range). By Belgian standards this is a prosperous farming area, by European standards even moreso. Field operations are as mechanized and efficient as anywhere in the world; dairying is much less so, partly because of the difficulty of adapting old buildings, but perhaps also because this tends to be regarded as "women's work"! The farm families themselves usually do not know their income levels precisely, as they usually do not keep complete accounts (income tax is levied on the basis of area). They are not poor, but neither are they well off by urban standards. They work long hours, seven days a week, and few of them ever take a holiday. Even the younger people rarely travel further than the nearest market town.

Table 1.2. Farm Structures[a]

	Number of Farms thousands		Average Size of Farms in hectares		in ESU[b]
	1960	1987	1960	1987	1985
Ireland	281	217	16.2	22.7	8.7
United Kingdom	443	243	32.0	68.9	42.0
Denmark	194	86	16.0	32.5	30.9
Netherlands	230	117	9.9	17.2	43.8
Belgium	199	79	8.2	17.3	23.8
Luxembourg	10	4	13.4	33.2	18.7
France	1773	912	17.0	30.7	22.8
Germany (BRD)	1385	671	9.3	17.6	17.3
Italy	2756	1974	6.8	7.7	7.9
Greece	878[c]	703	4.0[c]	5.3	-
Spain	3008[d]	1540[e]	14.8[d]	16.0[e]	5.9
Portugal	816[f]	384[g]	6.3[f]	8.3[g]	3.0

[a] Holdings of less than 1 ha. not included.
[b] The European size unit (ESU) measures the economic size of the holding. For each farm enterprise (wheat, cows, pigs), a "standard gross margin" is calculated on the basis of the average value of production minus the "specific costs" of production in the region in question over a three-year period. One ESU corresponds to a standard gross margin, thus calculated, of 1,000 ECU.
[c] 1961 (from *Agricultural Census*).
[d] 1962 (from *Anuario de Estadistica Agraria*).
[e] 1982 survey.
[f] 1968 (from *Inquerito às Exploraçoes Agricolas do Continente*).
[g] 1979/80 survey.

Sources: Eurostat, *Yearbook of Agricultural Statistics*, 1975 (Table C.1), and Commission, *The Agricultural Situation in the Community*, 1988 report (Table 3.5.4.2), and 1989 report (Table 3.5.4.1).

Structural change is occurring—the smaller farms, when they become vacant, get attached to larger ones—but it is difficult to imagine it happening any faster.

Output and Trade

Community farmers have been subject to much the same pressures as those in North America; Cochrane's "treadmill" theory applies in western Europe as well. Farmers have had to adopt new, output-increasing technology, or be left behind; our price-support system has, on the whole, encouraged them to do so, though many farmers have incurred excessive debt in the process (this has been a particular problem in Ireland and Denmark).

Output has gone up steadily, while food consumption, already at a high level, has not increased at nearly the same rate. Self-sufficiency rates have increased for almost every item (Table 1.5). Contrary to some interpretations, this has not been a deliberate EC policy aim, though some member states have encouraged the trend. Up to 1986, the EC's intervention agencies accumulated large stocks of cereals, dairy products, beef, and other items (Table 1.6).

Table 1.3. Farm Net Value Added (FNVA)[a] per Agricultural Work Unit (AWU)[b], 1987/88

	Average of all Farms	Percentage of Farms with New Value Added per AWU:	
		below 4,000 ECU	above 24,000 ECU
	000 ECU	percentages	
Ireland	10	28	5
United Kingdom	16	14	15
Denmark	19	29	22
Netherlands	25	10	46
Belgium	19	5	26
Luxembourg	15	8	18
France	14	12	14
Germany (BRD)	10	27	8
Italy	8	37	4
Greece	5	58	0.7
Spain	7	33	6
Portugal	3	81	0.3

[a] FNVA is the total value of output less intermediate consumption and depreciation, adjusted to take account of taxes, grants, and subsidies insofar as these are linked to production.

[b] An AWU represents the agricultural work done by one full-time worker in one year; the actual number of workers per farm is adjusted accordingly. Except on the larger farms, most of the AWU consists of the farmer and his family. FNVA per AWU corresponds to the remuneration of family and hired labor, own and borrowed capital, and management. To obtain family farm income per family work unit, wages, interest charges, rent, etc. must be deducted.

Source: Commission, *The Agricultural Situation in the Community*, 1989 report (Table 3.2.4), from Farm Accountancy Data Network.

Table 1.4. Farmers' Other Employment (1987)
(as percentages of the total number of farmers in each country)

	No other gainful employment	With other main gainful employment	With other secondary gainful employment
Ireland	63	26	11
United Kingdom	76	14	10
Denmark	67	10	22
Netherlands	76	15	8
Luxembourg	81	14	4
Belgium	67	30	3
France	68	12	20
Germany (BRD)	57	38	5
Italy	76	20	3
Greece	67	27	7
Spain	70	23	6
Portugal	61	32	6

Source: Commission, *The Agricultural Situation in the Community*, 1989 report (Table 3.5.1.6), from Eurostat "Surveys of the Structure of Agricultural Holdings."

Table 1.7 shows some broad consequences for agricultural trade. In the Community of Six, between 1963 and 1972, "intra-EC" trade increased rather fast, while imports from third countries rose only slightly (in value terms). Between 1973 and 1984, in the Community of Nine (or rather Ten, as Greece has got mixed up in these figures), the most significant factor was the large increase in EC *exports* to the rest of the world. The growth in output was, in effect, spilling over.

The major exception to this trend has been livestock feed (Table 1.8). Imports of both "cereal substitutes" (manioc from Thailand and other Far East countries, maize gluten feed and other products from the U.S.) and "protein-rich" feeds (mainly soya-based, from the U.S. and Brazil in particular) have increased significantly. The levy on manioc is "bound" under GATT at 6 percent; corn gluten feed, soya, etc. have zero bindings. Their price advantage has caused not only the cereal substitutes but also the soya-based feeds to be substituted for feed grain.

This is a sensitive matter in the GATT context, but it must also be stressed that there are strong feelings about it from the EC's point of view. The GATT bindings were granted in the early 1960s, when no one foresaw that trade in these items would become so important. Now, the

Table 1.5. Self-sufficiency in Foodstuffs
(production as percentage of internal use)

	Community of Ten		Community of Twelve
	>1973< or >1973/74<	>1984< or >1984/85<	1987/88 or 1987
Cereals (excl. rice)	91	118	111
Sugar	91	128	127
Wine	103	100	107
Fresh fruit (excl. citrus)	82	83	85
Citrus fruit	47	50	74
Fresh vegetables	94	101	106
Vegetable oils	-	47	63
Milk and milk products (fat content)	102[b]	119[b]*	110[b]*
Beef and veal	96[d]	108	107
Pig meat	100[d]	102	103
Poultry meat	102[d]	107	106
Sheep meat and goat meat	66[d]	76	81
Eggs	100[d]	102	102

>...< indicates a three-year average around the date shown.
a   >1984/85<.
b   Community of Nine.
c   >1984<.
d   >1974<.
*   Author's estimate.

Source: Commission, *The Agricultural Situation in the Community* (1989 report and earlier issues, various tables).

Table 1.6. EC Intervention Stocks (in thousand tonnes)

	1984 30/11	1985 30/11	1986 30/11	1987 30/9	1988 30/9	1989 30/9	1990 28/2
Butter	973	1018	1297	1058	221	30	52
Skim milk powder	773	514	847	722	14	5	5
Beef carcasses	468	589	452	484	559	180	130
Boned beef	127	214	220	207	164	60	-
Cereals (total)	9393	18647	12717	12235	10951	8920	9351
Olive oil	167	75	283	325	408	110	78
Alcohol (000 hl.)	-	501	666	1092	2892	3840	-

Source: Commission, *Proposals on the Prices for Agricultural Products and Related Measures (1990/91)*, COM (89)660 final (Table 6a). *Agra-Europe* of 16 March 1990 for end-February data.

Table 1.7. Trade in Agricultural Products[a] in the European Community

	Community of Six			Community of Ten			Community of Twelve	
	1963	1972	change	1973	1984	change	1986	1988
	million UA[b]		index	million ECU		index	mn. ECU	
Intra-EC trade	2,497	9,427	378	15,841	57,763	365	70,078	77,947
Imports from third countries	9,438	13,993	148	24,520	58,264	238	52,802	53,473
Exports to third countries	2,449	4,668	191	7,381	31,211	423	28,804	29,996

[a] SITC classes 0, 1, 21, 22, 232, 261-265 + 268, 29, 4, 592.1.
[b] Owing to changes in definition, 1972 data in units of account (UA) are slightly understated as compared with those for subsequent years in ECU. The ECU was introduced in the EC in 1979, replacing former units of account, and is used for the EC budget, common prices and subsidies under the CAP, and other purposes.
Source: Commission, *The Agricultural Situation in the Community* (various years).

Table 1.8. Community Imports from Third Countries of Products Intended Primarily for Animal Feed (in million tonnes)

	Community of Ten		Community of Twelve		
	1974	1980	1986	1987	1988
Manioc	2.1	4.9	5.8	7.0	7.0
Corn gluten feed	0.7	2.6	4.1	4.7	4.8
Other	2.4	5.5	5.1	6.6	6.8
All "cereal substitutes"	**5.2**	**13.0**	**15.0**	**18.3**	**18.6**
Soya beans[a]	7.2	9.3	10.4	11.6	
Soya cake	3.3	7.2	10.9	10.3	
Other seeds[a]	1.1	1.5	1.1	1.0	
Other cake	3.4	4.8	6.1	5.5	
Meat and fish meal	0.6	0.6	1.0	0.9	
Other	0.3	0.5	0.2	0.2	
All "protein-rich" products	**15.9**	**23.9**	**29.7**	**29.5**	

[a] In cake equivalent.
Source: Eurostat—COMEXT, published by commission in *The Agricultural Situation in the Community* (reports for 1986, 1987, and 1988, Tables 4.13.7.6). For 1974 and 1988, internal DGVI documents.

displacement of sales of EC-grown feed grain is seen as increasing the surplus for export to the world market, while a regional disequilibrium has occurred as intensive livestock enterprises (pigs and poultry especially) develop near the North Sea ports, to the detriment of less-favored regions.

A theoretical case for more balanced protection, based on Corden's "uniform tariff" doctrine, was put forward by Louis Mahé in 1984; Marloie (1985) showed the extent of displacement of production in France. The "Disharmonies" study sponsored by the EC executive Commission (1988) included as the basic option a 20 percent cut in cereal prices together with a 10 percent tariff on imported oilseeds. A recent paper in the French agricultural economics journal may also be quoted (Domecq, 1989—author's translation):

> An evaluation of the nutritive content of the imports of soya and cereal substitutes indicates that the Community of Ten would have to cultivate an area corresponding to the production of 100 million tonnes of wheat if it was to be self-sufficient in animal feed. The USA complain at EEC exports to the world market of 20 million tonnes of wheat . . . .
> Only some livestock producers, situated mainly in the countries bordering on the North Sea, have access to cheap imports of feed. The others, i.e., producers in the west and especially in the south of France, have to use European cereals for which they pay the European price. The result has been a divergent evolution of livestock production as between the north and the south of the Community . . . ."

Reform of the CAP

As mentioned previously, the word "reform" was resisted for a long time, particularly by the French; it was regarded as an "Anglo-Saxon" device aimed at undermining the "acquis communautaire" (established Community principles and practices), and contrary to what has been called above the "rhetoric" of the CAP—the reference to its "principles," and to its role as a "pillar" of the European Economic Community.

Nevertheless, much has changed over the last few years. The new measures will not be described here; other sources are available for this (e.g., Commission, 1988:1 and CAP Monitor). But it is necessary to explain how these changes came about.

The process was, in fact, gradual. The Commission—in particular the agriculture commissioner from 1973 to 1976, Pierre Lardinois—must take responsibility for some of the errors of that period, when prices were increased too much. But when Finn Olav Gundelach took over in 1977, he stressed the need for a "prudent" price policy; subsequently, the key word became "restrictive," and is currently "market-orientated."

By the early 1980s, the growing cost of agricultural price support to the EC budget (Table 1.9) could no longer be ignored. As has been pointed out, it aggravated the problem of the British budget contribution; further, the prospective accession of Spain and Portugal required an increase in the limit that had been placed on the total EC budget, and there could be no agreement on this unless spending on agricultural market support could somehow be curbed. These various items became the object of complex packages, which could only be resolved at the highest level—that of the European Council, composed of prime ministers or their equivalent (and in the case of France, the president).

The key decisions were taken (after long and difficult negotiations) by the European Council at sessions in Fontainebleau in June 1984 and in Brussels in February 1988. These decisions permitted a substantial increase in the total EC budget,[5] with a particular commitment to double spending on the "structural" funds (regional, social and the "guidance" section of the Community's Agricultural Guidance and Guarantee Fund—FEOGA) by 1993; they made possible the accession of Spain and Portugal (from January 1, 1986); and they included renewed budgetary concessions to the UK. An overall limit was set to spending on agricultural market support. In the context of this political process, the Agriculture Council adopted a number of specific measures.

The outcome for the agricultural sector is broadly as follows:

1. From 1988 through 1992, FEOGA "guarantee" expenditure (i.e., on market support) should not rise by more than 74 percent of the annual growth rate of EC GNP. The Commission's annual price proposals must be consistent with this limit, and "if the Commission considers that the outcome of the Council's decisions on these price proposals is likely to exceed the costs put forward in its original proposal, the final decision shall be referred to a special meeting of the Council attended by the Ministers of Finance and the Ministers of Agriculture which shall have the sole power to adopt a decision" (Conclusions of the European Council, February 19, 1988).

In the course of each budget year, an "early warning system" operates, under which the Commission, if it finds that monthly expenditure is running ahead of "expenditure profiles," uses its management powers to remedy the situation. If this is not sufficient, it presents proposals to the Council, which must act within two months.

Table 1.9. Community Expenditure on Agriculture
(in billion UA/EUA/ECU)[a]

	Community of Nine		Community of Ten	Community of Twelve				
	1973	1980	1985	1986	1987	1988	1989[b]	1990[b]
FEOGA "Guarantee" section (Titles 1 and 2—market support, etc.)	3.8	11.3	19.7	22.1	23.0	26.4	26.7	26.5
of which:								
– export refunds	1.1	5.1	6.6	7.2	9.1	9.5	8.9	9.1
– withdrawals from market and storage	0.4	1.9	5.3	6.3	4.8	4.8	5.1	4.9
– subsidies	1.9	3.4	7.7	8.0	8.8	11.1	11.7	11.8
Stock depreciation and special butter disposal	-	-	-	-	-	1.2	1.4	1.5
FEOGA "Guidance" section (structural reform, etc.)	0.2	0.5	0.7	0.8	0.8	1.1	1.4	1.6
Set-aside and income aid	-	-	-	-	-	-	0.1	0.3
All above items	4.0	11.8	20.4	22.9	23.8	28.7	29.6	29.9
Total Commission budget[c]	4.9	16.0	27.5	34.6	35.7	40.9	44.1	45.9

[a] For purposes of this table, changes in the definition of the unit of account can be ignored.
[b] Budget credits.
[c] Other EC institutions account for relatively small amounts (867 million ECU in 1990).

Source: Annual Community budgets, in *Official Journal of the European Communities*, and Commission, *The Agricultural Situation in the Community* (various years) (Table 3.4.4). Data refer to payment appropriations (not "commitments").

2. Since 1984, common prices have been frozen or reduced (Table 1.10). For cereals, oilseeds, and other commodities, cuts in prices or subsidies are now automatic under the "stabilizer" arrangements, if global output exceeds stated ceilings. Further, intervention

mechanisms for several products—in particular cereals, beef, butter and skim milk powder—have been substantially weakened.

These measures combined led to significant reductions in prices to producers, even in nominal terms, and still more when deflated by the cost-of-living index (Table 1.11); the relatively large price cuts to German farmers should be noted. In 1988 and 1989, prices recovered, largely due to world-wide factors (North American drought in particular), but downward pressures now appear to be recurring.

3. Milk output has been restricted by a quota system since 1984. So has sugar production, since the beginning of common market organization.
4. A "set-aside" scheme is in application, together with programs of "extensification" and "conversion."[6]
5. A program of direct aids to income ("decoupling," in GATT terms) is about to be implemented.[7]

Each of these measures can be criticized. The impact of cuts in common prices, for example, has sometimes been reduced for producers in countries with weak currencies by the "agri-monetary" arrangements. And the milk-quota system has given rise to many problems. The stabilizers, being related to global output, do not necessarily cause individual producers to reduce production. And the set-aside scheme, as U.S. experience has demonstrated, is fraught with difficulties. Finally, the direct-aids program seems complicated, and initially not much money has been allocated.

The point to make, however, is that, bearing in mind the difficulties of EC decision-making, and also the politically sensitive nature of agricultural policy, a remarkable amount has been achieved. And if we inquire *why* it has been achieved, the key point is that *further progress of the European Community demanded it*. A crisis had arisen, where spending on the CAP had reached unacceptable levels, preventing solutions of essential, interrelated issues, in particular the increase in the EC budget and the accession of Spain and Portugal.[8]

Personal factors sometimes also played a role. The Commission of 1985-88 was probably the most effective since the early days of the EC, and the political process already outlined owed much to its key members: Delors as president, Andriessen in the agriculture post, and Christoffersen with budget responsibility. Further, as was noted above, little can be done in CAP matters without France. Significantly, by 1984 it was becoming apparent that the accession of Spain and Portugal would shift France from being a net beneficiary under the EC budget to being a net contributor. The

Table 1.10. "Institutional" Prices—Selected Products (in ECU per tonne)

Marketing years	1973	1978	1983	1984	1985	1986	1987	1988	1989	1990[a]
Common wheat:										
– intervention price[b]	106	137	185	183	179	179	179	179	174	169[c]
– threshold price	113	159	256	254	250	251	251	246	237	
Sugar: intervention price (white sugar)	236	328	535	535	542	542	542	542	531	531
Rapeseed: intervention price	204	288	438	429	421	408	408	408	408	408
– applied[d]								373	394	?
Sunflower seed: intervention price	206	314	522	533	525	535	535	535	535	535
– applied[d]								419	499	?
Soya beans: guide price	-	322	562	570	576	576	558	558	558	558
– applied[d]								501	451	?
Milk: target price[e]	118	177	274	274	278	278	278	278	278	269
Beef cattle: guide price	862	1260	2071	2050	2050	2050	2050	2050	2050	2000

[a] Commission proposals for 1990/91.
[b] Less 3 percent co-responsibility levy from 1986/87, and additional 3 percent levy from 1988/89.
[c] Allowing for the automatic cut of 3 percent under the "stabilizer" arrangement.
[d] Under "stabilizer" arrangements: unlike cereals where the reductions are cumulative, the cuts for oilseeds are applied on a year-to-year basis. (Direct communication from commission.)
[e] In the case of milk, overrun of individual production quotas results in levies on producers, not reflected in the target price.

Source: *Agra-Europe, CAP Monitor* (section 17 on prices and commodity sections); commission proposals on the prices for agricultural products and on related measures (1990/91), COM(89)660 final.

Table 1.11. Indices of Producer Prices and Costs (1980 = 100)

	1985	1986	1987	1988	1989[a]
		Nominal indices			
Soft wheat—Germany	92	88	85	78	79
Feeding barley—UK	116	116	110	109	108
Maize—France	145	133	128	118	
Oilseeds—France	161	155	140		
Cattle—France	143	136	137	145	
Milk—Netherlands	118	115	116	126	127
Community of Ten:					
All products	144	146	147	151	162
of which:					
–Cereals and rice	133	136	134		
–Oilseeds	146	143	128		
–Cattle for slaughter	131	126	126		
–Milk	140	144	146		
Purchase prices of means of production	142	139	137	141	148
		Indices deflated by consumer price index			
All products:					
Germany	85	81	78	77	81
United Kingdom	88	86	84	80	80
France	90	88	83	82	85
Netherlands	94	87	86	86	91
Community of Ten	88	85	82	80	82

[a] Estimates based on preliminary and incomplete data (cf. Eurostat, *Rapid Reports—Agriculture*, 1990:1, and *Agricultural Prices*, 3/4 1989).

Source: Eurostat, *EC Agricultural Price Indices*, Series 5B, 1-1988.

first important step in the reform process—the decision in March 1984 to introduce milk quotas—owed much to the energetic chairmanship of the Agriculture Council by Michel Rocard, then French minister of agriculture.[9]

Stress has already been placed on the growing weight of the southern European countries; they have an evident interest in the increased EC budget for regional and structural measures, particularly when they fear increased competition in the post-1992 context. The German attitude is clearly vital, but ambiguous; though strongly in favor of the enlargement and reinforcement of the EC, they want the agricultural reform process kept to a minimum. For the UK, on the other hand, CAP reform was a precondition for accepting the other elements in the package.

What Next?

On the face of it, the CAP reform process has been rather successful. Surplus stocks have been significantly reduced (Table 1.6) by the restrictions on purchases into intervention, and by an admittedly expensive disposal program (in particular, cheap butter sales to the USSR); what remained has proved most useful in providing food aid to Poland and Rumania. The growth in the cost of market support has been contained (Table 1.9), partly, it is true, because of relatively firm world prices in dollars. CAP spending is now well within the budgetary limit.

All this could change, of course, if world commodity prices collapse, if the dollar falls, or if EC output breaks through the thresholds set under the stabilizer agreement. But for the next couple of years, the market and budget situation are not likely to compel further drastic action.

Nor is there strong pressure from public opinion in EC countries. Economists, of course, can point to the inefficiencies of agricultural policy, and the resulting costs for consumers and taxpayers. The Australians have been particularly active recently in demonstrating how much better off we would all be if we gave up agricultural protection, and no doubt their calculations are broadly correct (cf., for example, the Centre for International Economics, 1988). However, policy makers are not likely to be impressed unless they see political advantage from making the recommended changes. As things stand, these changes would annoy farmers considerably, and while the general public—if it notices—might perhaps approve, few votes are likely to be swung as a consequence. The following results from a recent public opinion survey ("Eurobarometer," Commission, 1988:2) are of interest:

1. "Think that aid to farming is a good thing": from 48 percent in France to 75 percent in Luxembourg—EC average 59 percent. On the other hand:
2. "We can no longer pay out large sums both as consumers and as taxpayers": from 42 percent in Denmark to 71 percent in France—EC average 62 percent.
3. "Buying from abroad may be better than relying on subsidies": from 7 percent in Denmark to 49 percent in Germany—EC average 34 percent.

4. "European agriculture should be protected from imports": from 40 percent in the Netherlands to 67 percent in France—EC average 51 percent.
5. "The European Community must defend its position as the world's second largest exporter of agricultural products": from 50 percent in Portugal to 84 percent in France—EC average 71 percent.

As can be seen, there are contradictions in these results, but no politician looking at the overall picture is likely to conclude that s/he can advance his/her career by embarking on a crusade for further CAP reform!

Other aspects of agriculture are now much higher on the political agenda: the quality of food in particular, following a series of health scares over matters as diverse as hormones in meat, salmonella in eggs and poultry, and listeria in cheese. The rural environment is also a concern. Indeed, perhaps the most interesting current development in the CAP is the gradual incorporation of environmental provisions in its structural policy (cf. the latest amendments to the basic structures regulation in Council Regulation No 3808/89, Official Journal No L 371 of 20.12.89). For the southern European regions, rural development is a priority. These are elements in a much broader approach to the role of agriculture, increasingly seen not just as the production of food, but as the vital element in the rural environment and society.[10] The Commission's price proposals for 1990/91 (Commission, 1989:2) envisage several innovations whereby market support would be "modulated" in favor of both small family farms that are "economically vulnerable," and other areas "suffering from natural handicaps or structural weaknesses."

Leaving the agricultural sector for a moment, we might note that the EC has some important items on its agenda for the next few years. Completion of the "single market" by the end of 1992 is going ahead—not just formally, in the sense of adopting the necessary regulations, but more fundamentally, in that the concept of free movement of persons, goods, and services has proved to have wide appeal to the business community and even to the public at large. European integration begins to mean something. European monetary union is a further step, more controversial, perhaps (it is not just Mrs. Thatcher who has doubts about it), but nevertheless with a lot of steam behind it.

And then, unexpectedly, there is Eastern Europe. Suddenly the EC has a new role to play, a role that is still being worked out. Indeed, we can find here once again the political significance that inspired the founders of the European Economic Community. It is not just a matter of providing aid, though that is important, and it is noteworthy that the Commission was given the task of coordinating the aid program to Poland and Hungary orchestrated by the "24" OECD nations. More fundamentally, the economic strength of the EC and the democratic traditions of its member states provide a model and a pole of attraction for the Eastern European countries emerging from totalitarian rule. The success of the EC in creating a common market—including an agricultural common market—is favorably compared with the dismal failure of COMECON. The EC has already concluded trade and cooperation agreements with most of the Eastern European countries (involving so far only minor concessions on agriculture); we must see how we can establish new relations more appropriate to their reformed political status. President Bush, last December in Brussels, called for "both a continued, perhaps even intensified, effort of the 12 to integrate, and a role for the EC as a magnet that draws the forces of reform forward in Eastern Europe" (*Financial Times*, December 22, 1989).

It remains to be seen how these other preoccupations will affect the EC's approach to the GATT negotiations, but at the very least, EC policy-makers will have several important matters on their minds (Vice-President Andriessen, having responsibility in the Commission for external relations, is dealing both with the Eastern European developments and with the GATT). More fundamentally, this is a time when EC solidarity is of vital importance. One practical consequence may be that, given the delicate issue of German reunification, other member states will be reluctant to put too much pressure on the Federal Republic where other matters are involved, and third countries should perhaps think twice before trying to drive wedges between the EC member states.

The EC's Position in the Uruguay Round

The EC is well aware of the need for a successful outcome to the Uruguay Round, and of the importance of agriculture in that context.

However, the initial U.S. "zero option" demands, and aspects of the Cairns Group proposals, have seemed to many in the EC like yet another attempt to undermine the CAP. The Mid-Term Agreement reached in April 1989 in Geneva seemed to offer a better prospect of going ahead, but the U.S. proposals of October 1989 made matters more difficult again, in particular by demanding that variable import levies be "prohibited" and that "administered price policies" and "income support policies linked to production or marketing" be "phased out." Specifically, the demand that "policies, other than border measures, that have resulted in or are designed to result in domestic prices higher than prices prevailing on the world market" should be *prohibited* after a ten-year transition period seems equivalent to demanding the abolition of the CAP. The Cairns Group proposals of November 1989 were not much better from the EC's point of view, in that they demanded a phasing-out of export subsidies over a ten-year period.

In December 1989 the EC put forward its negotiating position on agriculture, adopted after discussions in the Agriculture Council and the Foreign Council. In this, the EC recalled the aims of the Uruguay Round as stated in the Mid-Term Agreement: i.e., to establish a "fair and market-oriented agricultural trading system," and to provide for "substantial, progressive reductions in agricultural support and protection."

The Mid-Term Agreement accepted the concept of an "Aggregate Measure of Support"; the EC proposed a "Support Measurement Unit" (SMU), which would cover "all measures which have a real impact on the production decisions of farmers," and all the main temperate-zone products. Supply-control measures would somehow have to be quantified. The SMU would be based upon a fixed external reference price, to eliminate market and exchange rate fluctuations. The Community observed that the objective of reducing support and protection would be facilitated if there were international arrangements to keep world markets in balance, particularly through stock management (but EC members did not appear to make this a condition of agreement).

"The movement towards the reduction of support should be clear," but the EC did not at this stage put any figures on its proposal; commitments to reduce support could be taken initially for five years, with a review in the fourth year of the extent and rate of further reductions. The base year for calculating reductions would be 1986, and credit would be accorded for measures taken since the Punta del Este declaration in September that year. The extent of reduction would be related, in some way to be defined, to the situation on world markets.

The U.S. and the Cairns Group had proposed that nontariff measures, including variable levies, should be converted to tariffs, for subsequent phasing-out. Somewhat surprisingly, the EC made a step in this direction, stating its willingness to include "certain elements" of tariffication in the rules governing external protection. It envisaged that frontier protection would be implemented by means of a *fixed* element, to be reduced at a rate similar to that agreed upon for the SMU; and a *"corrective element,"* which would remain to absorb exchange rate variations and world market fluctuations, in so far as these go beyond certain agreed limits.

The EC's concession over tariffication was, however, conditional: a solution must be found to *its* demand for "rebalancing" of support and protection—i.e., with reference to cereal substitutes and oilseeds (but there was no longer any mention of a fats and oils tax). The EC's problem in this respect has been explained above.

These points may help to explain the EC's position as it stands at the end of March 1990. By the time of publication, the Uruguay Round negotiations will have moved on, and it would be dangerous to try to predict their outcome.

Notes

1. The chapters on specific countries in Neville-Rolfe's book (1984) provide much valuable material on France, Germany, Italy, and the UK.
2. These developments are discussed at much greater length in chapter 12 of the author's book (Tracy, 1989). On the significance of "rhetoric" in agricultural policy, see Bryden (1988).
3. A thorough analysis of German agrarian politics was made by Gisela Hendriks in a doctoral thesis for Bristol University in 1985 entitled, "A Critical Analysis of West Germany's Approach to European Integration: the CAP—a Central Area of Conflict." Although the thesis is unpublished, the main points can be found in a subsequent article (Hendriks, 1987).
4. Much valuable information on the behavior of farm families, including resort to off-farm activities, is being provided by an ongoing multinational and multiannual survey conducted by the Arkleton Trust. One of the French members of this team, André Brun, has recently drawn attention to the "paradoxical logic" according to which many farm families seem willing to subsidize their farm activities by means of their other activities: "The family strategy is apparently to maintain the rural residence, the agricultural activity and the inherited property by means of income which does not depend on this source, in particular through the wages earned by one or more of the family" (Brun, 1989, author's translation).
5. The total contribution by any member state to the EC budget was previously limited to 1 percent of its "VAT base"—i.e., the total value of the goods and services which, according to a harmonized EC list, are liable for VAT. This ceiling was raised to 1.4 percent on the accession of Spain and Portugal in 1986, but again the total proved inadequate. The decision at the European Council of February 1988 introduced an overall ceiling of 1.2 percent of GNP (gross national product) for "payment appropriations" (1.3 percent for "commitment appropriations"). This represents a higher total than the VAT key.
6. These are provided for in successive amendments to the basic "structures" regulation No 797/85.
7. See Council Regulation No 768/89 of 21 March 1989 (Official Journal No L 84 of 39.3.89), and Commission Regulation No 3813/89 of 19 December 1989 (Official Journal No L 371 of 20.12.89).
8. That difficult decisions should be reached only in a crisis situation is a concept wholly in line with the analysis of political scientists—cf. in particular Moyer and Josling (1990).

9. See Petit et al. (1987) for a comprehensive review of the 1984 decisions. As this work suggests, Rocard's determination to push through milk quotas may reflect not only a shift in the perception of the French national interest; he seized an opportunity to make his personal mark after previous summits had failed. This may illustrate a factor of growing significance: each president of the council, in his six-month stint, nowadays wants some achievement to his credit.

10. Cf. the Commission's "Rural World" report of 1988; there are also perceptive comments in papers by John Bryden, 1988 and 1989.

References

Agra-Europe (weekly bulletin). London.
Agra-Europe, CAP Monitor (continuously updated). London.
Bryden, J.M. "European Rural Policy: Are Structural Changes in Agriculture Consistent with, or Contrary to, the New Goals of Rural Policy?" Paper to Forschungsgesellschaft fur Agrarpolitik und Agrarsoziologie, 10 Okt. 1989.
――――. "Some Recent Changes in the Underlying Political Basis of Europe's Rural Policies." Paper for the Symposium on Comparative National Perceptions and Political Significance of Rural Areas. Economic Research Service, USDA, Penn State University, and the Aspen Institute for Humanistic Studies, 1988.
Commission of the European Communities:
The Agricultural Situation in the Community (annual).
"Perspectives for the Common Agricultural Policy." Green Europe Newsflash no. 33, 1985.
"Changes to the EEC Market Organization for Milk and Milk Products." Green Europe Newsletter no. 220, 1987.
Implementation of Agricultural Stabilizers. COM(87)452 final, 1987.
"Restoring Equilibrium in the Agricultural Markets." Green Europe, 1988:1.
Europeans and Their Agriculture. Eurobarometer, 1988:2.
"The Future of Rural Society." Bulletin, suppl. 1988:3.
Disharmonies in EC and US Agricultural Policies. 1988:4.
A Common Agricultural Policy for the 1990s. European Documentation, 1989:1.
Proposals on the Prices of Agricultural Products and on Related Measures (1990/91). COM(89)660 final, 1989:2.
Domecq, J.-P. "Origine des Excédents Agricoles et leurs Rôles dans l'Économie française." *Economie Rurale* 194 (Nov.- Déc.1988), pp. 29-33.
Hendriks, Gisela. "The Politics of Food: The Case of FR Germany." *Food Policy*, Feb. 1987.
Macroeconomic Consequences of Farm Support Policies. Canberra: Centre for International Economics, 1988.
Mahé, L. "A Lower but More Balanced Protection for European Agriculture." *European Review of Agricultural Economics* 11 (1984), pp. 217-34.
Marloie, M. *Le rôle des Transports dans la Concurrence sur les Marchés des Céréales, des Oléagineux et des Aliments de Bétail.* Montpellier: INRA-ENSAM, 1985.
Moyer, W. and Josling, T.E. *The Politics of Agricultural Reform in the United States and the European Community.* London: Harvester-Wheatsheaf, forthcoming 1990.

Neville-Rolfe, E. *The Politics of Agriculture in the European Community.* London: Policy Studies Institute, 1984.

Normile, Mary A. *The European Community in the Uruguay Round: Agricultural Trade Negotiations and EC Policy.* USDA, Agriculture and Trade Report RS 89 2, 1989.

OECD. *National Policies and Agricultural Trade.* Paris, 1987.

Petit, M. et al. *Agricultural Policy Formation in the European Community: the Birth of Milk Quotas and CAP Reform.* Amsterdam, Oxford, New York, Tokyo: Elsevier, 1987.

Tracy, M. *Government and Agriculture in Western Europe.* London: Harvester-Wheatsheaf, and New York: N.Y. University Press, 1989.

Tyers, R. and K. Anderson. "Liberalizing OECD Agricultural Policies in the Uruguay Round: Effects on Trade and Welfare." *Journal of Agricultural Economics,* 39.2 (1988), pp. 197-216.

2

The Political Economy of Agriculture in Canada

Grace Skogstad

Introduction

As we enter the 1990s, Canadian agricultural policy appears destined to change. At home the catalysts to change are numerous and include a shift in government priorities, which has put fiscal restraint and budget deficit reduction at the top of the federal Progressive Conservative government's agenda; a period of prolonged economic distress in the grains sector, which has demonstrated the inadequacy of existing measures of support; a more than decade-long failure by domestic poultry marketing boards to put to rest their critics' charges, which include allegations of inefficiency and market balkanization; and a diminution in the strength of forces supporting existing agricultural programs as a result of the fragmentation of the farm lobby and the ascendancy of producer spokespersons for a more market-driven agriculture. Further pressures have come from abroad. Trade disputes with the U.S. and GATT panel rulings in the context of the multilateral effort to reform the world agricultural trading regime have imperilled traditional support measures of supply management and income stabilization. Domestic and international economic and political factors are thus pressing for reduced government spending on behalf of agriculture; at the same time they are casting in doubt the legitimacy and effectiveness of particular policy instruments.

But even while change seems inevitable, its direction and magnitude will be circumscribed by the political process within which agricultural policy in Canada is made and implemented. The process is one that makes some agricultural programs more entrenched and more resistant to change than others. To illustrate this, the chapter begins with a general overview of the agricultural policy process. It then identifies three strands to Canadian agricultural policy, explaining each in terms of the political economy of the commodities to which it applies. Armed thus with a description of the dynamics that underlie existing Canadian agricultural programs and initiatives, the chapter then considers the durability of the latter, probing into the altered domestic and international context and changed policy-making process within which agricultural policy is being made in the early 1990s.

The Agricultural Policy Process

Existing Canadian agricultural programs have evolved out of a policy network that can best be described as a relatively closed one dominated by three principal parties: the national government, provincial departments of agriculture, and producer organizations. Although Ottawa shares jurisdiction with the provinces for agriculture, at least until the 1970s it predominated not only by virtue of its larger purse, but also because provinces were willing that it do so, perceiving Ottawa to be primarily responsible for agricultural spending.[1] Federal dominance was enhanced by the national government's jurisdiction over interprovincial and export trade, which left it in charge of regulating the national grain handling and railway transportation systems, determining freight rate schedules, and with control over import tariffs and other border-control measures.

Provincial governments have always been important partners in this policy-making network. Provincial input into federal agricultural policy is institutionalized in various ways, including the annual agriculture ministers' conference. But more important than this ministerial forum is the very elaborate and extensive bureaucratic machinery that co-ordinates the multiple joint programs of the two levels of government. Federal and provincial officials are brought into constant contact as they implement agricultural research, crop insurance, and other programs. Provincial governments have also been policy initiators in their own right. They have exercised their legal authority over marketing within their borders to enable the creation of producer-controlled marketing boards, and in the 1970s used their shared jurisdiction over agriculture more aggressively to establish price and/or income stabilization programs. Although their greater policy activism has occasioned intergovernmental conflict across both levels of government, on the whole, federal-provincial relations in agricultural policy are harmonious and civil.

The third important party to agricultural policy-making is producer groups. The right for farmers to be consulted on matters of agricultural policy, historically facilitated by the early emergence of producer organizations, has long been recognized. It is a right which, in Canada's federal system, often gives producers two "kicks at the can"—the first by virtue of their own lobbying of federal or provincial decision-makers, and the second as a result of their skill in persuading provincial officials to make local producers' causes provincial causes before federal authorities.

Although most matters of agricultural policy entail a role for these three parties, their respective influence varies appreciably depending upon the aspect of agricultural policy under question. Provincial influence is greater where provinces have jurisdictional leverage; that is, where the policy matter is one over which the provinces share jurisdiction (commodity price stabilization) or exercise exclusive legal authority (intraprovincial marketing), as contrasted to one over which the national government has exclusive jurisdiction (tariffs, export trade). Producer influence is analogously variable and largely a function of first, the organizational unity and coherence of commodity producers; second, the electoral significance of the farmer vote; third, the capacity for producer-provincial government alliances against the federal government; and fourth, the economic significance of the commodity sector.

The next section briefly describes the major agricultural programs in Canada, noting the political and economic factors that gave rise to them, and the policy-making process that surrounded their formulation. The section begins with a discussion of the goals underlying Canadian agricultural policy.

Canadian Agricultural Programs and Goals[2]

Despite its often ad hoc and largely reactive nature, Canadian agricultural policy displays certain continuities. First, agricultural policy cannot be divorced from broader macroeconomic and national policies. Whether in the nation-building, post-Confederation period, the economic crisis of the Depression, the political emergency of the Second World War, the political instability of the 1960s, or the economic recession of the early 1980s, agriculture has received attention in proportion to its ability to further nonsectoral goals of political unification and regional economic development.[3] As will be discussed in further detail later, the impact of macroeconomic factors is no less real today. Second, and partly as a consequence of agriculture's wider importance, the state has involved itself in agriculture. This has included conferring upon farmers the means to establish their countervailing power in the market place, as well as government expenditures and regulations to overcome the imperfections of the market place. Third, government involvement in agriculture has, for the most part, been directed to the goal of enabling farmers to become more productive, more efficient, and hence, more competitive. The major instruments to achieve an efficient and competitive agricultural industry have included federal government funding for scientific research, provincial education and extension services, eased access to agricultural credit (the Farm Improvement Loans, the Farm Credit Corporation), and an aggressive search for foreign markets.

Fourth, although government support has not, with certain important exceptions, been intended to replace the market place as the source of farm incomes, beginning in the 1940s, and more so since the late 1950s, governments have defined their responsibility to include protecting farmers' incomes at an adequate and stable level. The provision of government low-slung "safety nets" grew out of the recognition that producer incomes are determined by climatic and other factors beyond the control of farmers, and a serious lag in farm incomes behind industrial wage incomes by the 1950s. Fifth, rather than pursuing a uniform approach across all sectors of agriculture, governments have adopted a commodity approach, making programs and policies on a commodity-by-commodity basis, as the economic characteristics and political goals of the sector have dictated. Thus, the extent and type of government involvement varies appreciably from sector to sector. It also varies across provinces, owing to divergences in producers' (and their provincial governments') preference for collective action and government involvement in agriculture. Sixth, the goal of balanced agricultural development across the regions of Canada has figured somewhat in national agricultural policy (the Feed Freight Assistance Act being the most visible evidence), but it has been left largely to the provinces to assist the production of regionally specific commodities and to offset comparative disadvantages.

If these objectives are relatively clear, others are less so. Is the preservation of the family farm an objective of Canadian agricultural policy? Some farm groups, pointing to the consistent decline in the number of farm units and the emphasis on a capital-intensive and large-scale agriculture, say it is not and never has been a goal. That the family farm remains the predominant unit of production is no doubt at least indirectly the consequence of government policy. The emphasis has been on the economically efficient family farm, not as a goal in itself, but rather as a means to higher goals, including adequate incomes.[4] In this sense there is a noticeable difference between North America and Europe.

Another cross-national difference of goals centers on the objective of food security. Again, the assurance of safe and secure food supplies always figures in governmental listings of agricultural goals. But, unlike Europe and Japan, Canada has never experienced a lack of food security. War did not jeopardize our food supplies, which have always been plentiful by virtue of our own productive capacities and our easy access to external sources to supply foodstuffs in which we are deficient. In this sense, the goal of food security is largely symbolic and lacks the emotional and experiential content that it carries in Europe and Japan.

But if food security has been more symbolic than real, the same cannot be said for the goals of safe food supplies at reasonable prices. Governments have taken both goals seriously, as witnessed by the plethora of health and protection standards that regulate food quality, and the commitment to production and marketing efficiency to ensure low-cost food.

Although the goals of producer income security and stability and of production efficiency transcend agricultural policy generally, differences in the political/economic structures of various commodity sectors have resulted in different policy instruments being used to achieve these goals. Viewed collectively, the programs and instruments across sectors are contradictory, and they give rise to the observation that Canada's most important agricultural commodities are subject to three different policy strands: first, protectionism/market regulation, which characterizes the dairy and poultry sector;[5] second, state assistance, which prevails in the grains and oilseeds sector; and third, market orientation, which dominates in the hog and cattle sectors. While these strands characterize each commodity sector to which they apply, their dominance varies to some extent across provinces.

Protectionism/Market Regulation

Dairy, egg, and poultry (chicken, turkey, and broiler/hatcher) producers sell their commodities in the Canadian market protected from external competition, and their incomes are determined by government subsidies (dairy) and/or government-approved pricing formulae (dairy, poultry).[6] The instruments that achieve market protection and income guarantees are a national system of supply management that establishes provincial and national production/marketing quotas, import controls (legitimated by GATT Article X1.2.c when domestic production is managed), formula pricing (which guarantees production costs), and, in the case of the dairy sector, government subsidies. Collectively, these policy instruments provide dairy and poultry producers with a degree of income security and stability far greater than that enjoyed by any other agricultural sector.

Explanations for this strand of Canadian agricultural policy lie in the political economy of the late 1960s and early 1970s. The creation of the Canadian Dairy Commission (CDC) in 1966 and the implementation of a nation-wide program of supply management, federal support payments, and administered prices in the early 1970s, were a response to militant producer mobilization in the wake of a severe economic decline and chaos in the industry. The political economy of the dairy sector was such as to compel both provincial and federal attention. The industry is concentrated in Central Canada (today the region produces 78 percent of industrial milk and 64 percent of fluid milk).[7] As Table 2.1 shows, dairy sales and payments account for 36 percent of Quebec farm cash receipts and about 20 percent of Ontario farm cash receipts. Both producers and milk processors were highly organized[8] and when fifteen to twenty thousand Ontario and Quebec farmers mounted "the largest demonstration ever to take place on Parliament Hill" to demand higher industrial milk prices and salvation of their industry, the minority federal Liberal government was extremely vulnerable.[9] Recognizing that the support of the alienated rural Quebec voter would restore its majority status, the Liberal government was anxious to assuage dairy farmers' anxieties. It was able to do so because provincial governments in Ontario and Quebec, and eventually everywhere else in Canada, agreed to cooperate in sharing the Canadian market for fluid milk products, by limiting production within their own provinces. In return for provincial cooperation in an area constitutionally off-limits to it (intraprovincial marketing), the federal government utilized its exclusive powers to regulate the extraprovincial and export flow of dairy products, and drew on its deeper purse to subsidize milk and cheese prices.

Similar economic and political circumstances account for the creation of supply management schemes in the poultry sector. Depressed poultry and egg prices, leading to cut-throat prices, and followed by "wars" to keep out other provinces' products, were only resolved by the Supreme Court's affirmation that provincial marketing boards could not restrict interprovincial trade by preventing the entry of products from another province.[10] But the provincial and producer cooperation that facilitated resolution of problems in the dairy sector was not as readily forthcoming. Remedial federal legislation to enable the creation of national producer-controlled marketing boards was bitterly opposed by the opposition Progressive Conservative Party and the provincial governments of Alberta and Saskatchewan, who successfully allied themselves with red-meat commodity groups to procure the latter's exemption from the Farm Products Marketing Agencies Act

Table 2.1 Provincial Sources of Farm Cash Receipts by Commodity, 1987 (%)

	Nfld.	P.E.I.	N.S.	N.B.	Que.	Ont.	Man.	Sask.	Alta.	B.C.	Canada
Wheat, Oats, Barley	-	1.2	.5	1.0	1.4	1.5	22.7	40.0	20.3	1.5	14.9
Crop Insurance Payments	.1	1.5	.2	1.9	1.6	.3	1.6	2.6	3.3	.9	1.7
WGSA[a]	-	-	-	-	-	-	12.5	17.9	9.3	.8	6.7
Oilseeds[b]	-	-	-	-	-	-	8.4	7.9	7.8	.4	4.0
Potatoes	2.5	3.8	2.5	21.5	1.5	.9	2.1	.2	.9	1.9	1.7
Fruits and Vegetables	6.1	3.2	10.6	7.3	5.4	8.8	.8	0.0	.8	18.2	4.6
Cattle and Calves	2.4	13.7	11.3	9.6	9.6	19.5	15.5	13.4	30.6	17.4	18.0
Hogs	7.9	12.4	11.6	8.2	20.1	12.4	12.2	2.7	7.2	4.6	10.2
Dairy[c]	24.9	17.3	29.1	23.5	36.3	19.8	5.8	2.3	6.1	22.7	15.1
Poultry	26.7	.5	11.5	9.6	8.3	6.5	2.4	.8	2.4	10.4	4.7
Eggs	17.9	1.4	6.9	5.6	2.8	3.3	2.3	.4	1.1	5.5	2.3
Prov. income stabilization and supplementary payments	3.4	1.8	.4	.3	2.7	2.3	7.5	9.8	6.9	.8	5.1

Do not add to 100 because of the omission from the table of less nationally/provincially significant commodities.
[a]Western Grain Stabilization Act payments   [b]Rye, flaxseed, canola/rapeseed   [c]Including supplementary payments
Source: *Handbook of Selected Agricultural Statistics*, Table 9, p. 7.

(FPMA).[11] The compromises the federal government was forced to make to secure passage of the FPMA saddled the poultry agencies subsequently set up under its authority (the Canadian Egg Marketing Agency, the Canadian Chicken Marketing Agency, and the Canadian Turkey Marketing Agency) with a decision-making structure that has caused much internal wrangling among provincial producer representatives on the national marketing agencies, and recurrent disputes between marketing agencies and their national supervisor (the National Farm Products Marketing Council).[12] Subject to repeated charges of ignoring consumer interest, of fostering inefficiency, and of being unaccountable to governments, poultry agencies have fought an uphill battle to establish their legitimacy.[13]

State Assistance

Producers of grains and oilseeds, located in the largest agricultural region of Canada, are governed by this second strand of Canadian agricultural policy. Although it entails a reliance upon international markets to determine commodity prices, income vulnerability is cushioned by two items: the provision of government financial assistance to stabilize prices and incomes, and government regulations to enable producers to enhance their collective bargaining power.

Economic realities dictate the market orientation thrust; Canadian grain and oilseeds producers are price-takers in an external market upon which they must rely to absorb the largest portion of their crop. State assistance to the grains and oilseeds sectors is rooted in their sheer economic importance to the three prairie economies and the Canadian economy as a whole. (See Table 2.1.) Economic downturns in the prices of these commodities have been a catalyst to the creation of particular programs, each of which arose out of discrete political circumstances: sometimes because of effective political representation in the federal government and cabinet, sometimes in the face of a politically vulnerable national government, but almost always as the result of political pressure from united provincial governments and well-organized producer groups.

The major programs of state assistance are as follows: subsidization of railway freight rates for the shipment of grains, from 1897 to 1982 by means of the Crow's Nest freight rates and since then by means of federal payments to the railways under the Western Grain Transportation Act; federal regulation of the grain-handling system; the creation of the Canadian Wheat Board, a producer-controlled marketing agency to augment producers' collective marketing power, which ensures equity in marketing opportunities, and, with government guarantees of initial payments, provides a floor on grain prices; stabilization of the returns from the sale of major crops in the prairie region, through the Western Grain Stabilization Act; and protection from crop loss via the federal-provincial shared cost Crop Insurance Act, and periodic disaster-relief payments, such as the Special Canadian Grains Program (1986-88).[14]

The national government's regulation of grain marketing and handling, like that of grain freight rates, occurred during a period in which farmers constituted a significant electoral force, both provincially and nationally. The creation of the Canadian Wheat Board in 1935 and its monopoly over domestic and export marketing of wheat, oats, and barley between 1947 and 1973, was a reaction initially to the emergencies of depression and war, as well as to sustained "direct and indirect political pressure" from provincial legislatures and grain farmers.[15] Influential western spokesmen in federal cabinets (in particular, Jimmy Gardiner in the Liberal government of Louis St. Laurent and Alvin Hamilton in the Conservative government of John Diefenbaker) were also important in explaining these policies, as they were in accounting for the durability of the Crow's Nest freight rates for over eighty years, and the implementation of crop insurance in 1959.[16]

The decline in the electoral influence of the farm vote was offset to an important extent by cohesive grain growers' organizations, which were able to ally themselves with provincial governments sympathetic to their goals. Thus, when economic circumstances lent momentum to producers' pressure for a more effective program to stabilize western grain returns, the unrelenting pressure of the western provincial governments, farm groups, and opposition parties in the House of Commons, proved instrumental in effecting important changes to what became the Western Grain Stabilization Act.

The passage of the act, which terminated the Crow's Nest freight rates, but retained government subsidization of grain transportation, provided a reverse example of the importance of a cohesive farm lobby and a united provincial stance in affecting agricultural policy in areas of jurisdiction under exclusive authority. In the absence of unity among Western Canadian producers and their provincial counterparts—that is, with divergent and conflicting goals among western grain and livestock producers and provincial governments—the forces that prevailed in national grain transportation policy were those that were able to maximize their electoral significance by aligning themselves with a cohesive Quebec farm-processor lobby adroit at lobbying its provincial government and rural representatives in the federal caucus.[17]

Market Orientation

Policies for the red-meat sector illustrate the third strand of Canadian agricultural policy: an overall preference for and reliance upon the market place to determine incomes. This sector has few import barriers and those technical and tariff barriers that exist will be largely eliminated as the Free Trade Agreement (FTA) is implemented. The Meat Import Act enables, but does not require, the federal government to impose quotas on beef and veal imports on a countercyclical basis. The principal and only significant program of state assistance is that from which almost all other Canadian farmers benefit: the income safety net of the Agricultural Stabilization Act.

Economic factors help to explain the market orientation of cattle and hog producers; they operate within a North American market in which prices are largely determined outside the country. Ideological factors are also important; the chief organizations representing this sector, the Canadian Cattlemen's Association and the Canadian Pork Producers' Association, are philosophically committed to a laissez-faire market.

In spite of this general bias toward the market, other political/economic factors intervene to result in considerable interprovincial variations in state assistance to red-meat producers (and so to render the label "market orientation" somewhat of a misnomer). Policy toward the red-meats sector has been affected by federalism, that is, by the province's shared jurisdiction for agriculture. The latter has enabled provinces committed to an activist state to spend to offset economic differences in costs of production and comparative advantage. Selected provinces (B.C., Quebec, Ontario) moved in the 1970s to prop up producer incomes when the federal government was loathe to enhance its support. Their actions mobilized producers in provinces that had not chosen to "top load" federal price support. As a result of concerted pressure by the Canadian Cattlemen's Association and western provincial governments, and protracted negotiations that have seen provincial and national governments working closely with cattle- and hog-producer organizations, changes in the Agricultural Stabilization Act have achieved a degree of harmonization between the federal and provincial programs.[18] Funding is now tripartite—producers, provincial governments, and the federal government each pay premiums—and limits exist on the ability of provinces to offset regional differences in cost of production. But the very recognition of the legitimacy of some degree of provincial "top loading" affirms the influence provincial governments and producer organizations—in this case, those centered principally in Quebec—can have on national agricultural policy.

The political economy of Canadian agriculture creates enormous potential for conflict between producers in different provinces and between producers of different commodities in the same region. The tendency toward specialization of commodity production within a province (Ontario is the exception) shown in Table 2.1, when accompanied by provincial jurisdictional authority over agriculture, enhances the probability of legislation that protects local producers at the expense of producers in another province. But because there are philosophical differences across provinces in the perceived appropriateness of government involvement in agriculture, provincial government intervention will be uneven. The result will be differential treatment of producers across the country and ensuing interprovincial conflicts. Conflict also arises because the interdependence of producers of different commodities—such that high prices for grain growers mean high costs for livestock producers—heightens the likelihood that government intervention on behalf of one commodity group will have negative repercussions for another.

To summarize, strong producer organizations and activist provincial governments have shaped Canadian agricultural policy. Producer and provincial roles and influence on each of the three strands, however, are somewhat different. Three distinct policy networks—that is, three different sets of relationships among producer groups and governments—underlie each of the three policy orientations. The network that describes supply management/market regulation is one which institutionalizes producer and provincial input into federal policy. Provincial cooperation is required to create national marketing agencies, and any amendments to their marketing plans require provincial consent. Poultry and dairy producers must also give their consent to the creation of marketing agencies, and their representatives help to implement the marketing plans (under the supervision of government regulators). In terms of their functioning, these policy networks are labelled corporatist;[19] dairy and poultry producers, processors of these commodities, and government officials, collectively administer the dairy and poultry marketing plans. In these relatively closed networks of mutual interdependence, the federal government is not without a trump card. It alone has sole authority to put in place or to remove controls on competitive imports, and will do either, depending upon its political calculus (the significance of the Central Canadian farm vote) and philosophical bias. And in the case of the dairy industry, its leverage also includes subsidies.

The policy network that sustains the other strand, which entails appreciable state involvement but without market regulation—that in the grains and oilseed sectors—is one in which producer groups and federal authorities are the key actors. Rather than being active participants in the implementation of sectoral policy, producer groups here are essentially lobbyists, advocating policies and pressing for action from outside the decision-making circles. Their task is more difficult because of organizational pluralism: there is not one association that purports to represent producer interests, but several, competing groups, some of which (for example, the Western Wheat Growers' Association and the Saskatchewan Wheat Pool) advocate diametrically opposed measures. Without jurisdiction in this area, the capacity of provincial governments to shape policy outcomes is much diminished, and largely a function of the personal influence and credibility of provincial ministers and premiers. The policy network here is more open and pluralist, and the federal government has latitude for independent action, depending upon the strength of producer groups, electoral imperatives, the presence in the government of effective regional political representatives, and the economic significance of the grains and oilseeds commodities.

Pressure pluralism[20] also characterizes the policy networks that gird the market orientation thrust in the red-meats sector. Producer groups, provincial governments (by virtue of their shared jurisdiction for agriculture), and the national government are all significant actors. As documented above, provincial activism can produce fragmented and conflictual policies. With provincial governments serving as potential veto points, efforts to effect policy change can be thwarted. At the same time, however, producer-provincial alliances can be potent weapons in seeking change in federal policy. A case in point are the recent subsidies provided by the four Western Canadian provinces to local livestock producers to offset the inflated cost of prairie feed grain, a result of Western Grain Transportation subsidies to the railways. Such provincial initiatives are obviously a bargaining ploy of support for local livestock growers seeking to reverse the current federal policy of paying grain freight subsidies to the railways, rather than to the producers.

Policy-Making in the 1990s

In the latter part of the 1980s and early 1990s, a number of factors have destabilized the policy networks that underlie current agricultural programs. Most destabilizing have been developments in the international arena. The breakdown of the international trading regime has witnessed EC-U.S. grain trade wars and heightened U.S. protectionism, which has intensified trade disputes between Canada and the U.S. These disruptive events, in turn, have precipitated a four-year round of the GATT to restore order and liberalism to world trading relations. Parties to the multilateral negotiations (MTN) have committed themselves to reduce substantially agricultural support and protection by eliminating or capping export subsidies, and by reducing the trade-distorting effects of domestic subsidies.[21]

International occurrences have set off their own chain of events in Canada. EC-U.S. grain trade wars contributed to an economic recession in the grains industry, which, in turn, prompted an unprecedented level of federal financial support in the form of the Canada Special Grains Program.[22] The penetration of international trade issues into domestic policy led to a heightened attention in government circles to trade matters, and a priority on negotiating new trading arrangements, first, with the U.S. in the form of the bilateral FTA, and second, in the multilateral forum of the GATT. As will be discussed in greater detail, the bilateral and multilateral negotiations readjusted the relationships among producers and provincial and national governments, as the government of Canada created new domestic political institutions to formulate and negotiate trade agreements. They also put Canada's agricultural policy under the microscope of international scrutiny, which, in the wake of the U.S.-initiated GATT ruling pursuant to the FTA on the legality of import controls on further-processed dairy products (yoghurt and ice cream), has undermined the legitimacy of the policy instrument of supply management. Trade remedy dispute rulings in the U.S. concerning Canadian exports of hogs and pork have similarly cast doubt on the appropriateness of government-subsidized price stabilization programs. And indirectly, the very efficacy of state intervention in the grains industry has come to be questioned, as domestic safety nets have proved to be an inadequate defence against the treasuries of foreign countries. In short, the policies and trade rulings of foreign countries have caused international trading concerns to be injected into domestic agricultural policy to an unprecedented degree, have led to an alteration in the decision-making process, and have rendered vulnerable traditional policy instruments of support.

At the same time, domestic developments independent of these international forces also threaten to disrupt the policy networks sustaining current agricultural programs. Three developments are especially important: first, a shift in the composition of the national government, and with it a change in goals; second, changes in the domestic policy process, which broaden the policy community with input into agricultural matters; and third, a fragmentation of the farm lobby and a consequent diminished capacity to mount a united producer front.

In 1984 the Progressive Conservatives replaced a government without significant representation in the prairie farm community with one which had able regional representatives in the federal caucus and cabinet. With overwhelming support in the rural ridings of Alberta and Manitoba, and in a very competitive situation in rural Saskatchewan, the federal Conservative government has considerable legitimacy to speak on agricultural issues pertaining to Western Canada. Producer trust was further bolstered by the Conservative government's readiness to provide unprecedented levels of financial support for hard-pressed prairie grain farmers (the Canadian Special Grains Program) between 1986 and 1988. Complementing this prairie support is the overwhelming electoral endorsement of rural Quebec and Ontario in 1988.

The shift in the partisan complexion of the national government has also been accompanied by the injection of a new set of priorities into national policy making. These are a heightened priority on deficit reduction and fiscal restraint, the goal of national reconciliation through more cordial and collaborative relations with provincial governments, and a commitment to a market-driven economy with less state intrusion.[23] The goals of national reconciliation and deficit reduction mesh as Ottawa seeks, with some success, to have provinces assume greater financial responsibility for agriculture. Together, the strength of the Western Canadian caucus and the laissez-faire bias imply more limited support for supply management, especially when one recalls the resistance of western opposition Conservatives to the passage of the FPMA Act (which enables supply management plans) in 1971-72.

The second domestic development with implications for changes in agricultural policy networks has been a further acceleration in the late 1980s of the shift from departmentalized decision-making to institutionalized cabinet decision-making.[24] For the past two decades the autonomy of departments, including Agriculture Canada, over decision-making has been eroded as cabinet committees have become the important locus of expenditure and policy decisions. Within cabinet committees, ministers of other departments collectively decide on the appropriateness of government initiatives regarding agriculture. The Mulroney government has further centralized decisions regarding new expenditures in a small cabinet committee composed of the most senior ministers, including the finance minister. It is instructive to note that the minister of agriculture, also the deputy prime minister, is vice-chairman of this Expenditure Review Committee—his word on agricultural spending thus carries enormous weight. As a whole, the centralization of decision-making and interdepartmental coordination has augmented the capacity of the national government to formulate coherent policies.

Simultaneous with this narrowing and centralization of financial decision-making has been a broadening of the policy community within which agricultural policy is made. The emphasis on cordial federal-provincial relations has increased opportunities for provincial input into federal policy, and led to closer federal-provincial relations. The same is true of producer consultation; government-producer task forces proliferate. The most important formalization of provincial and producer input has been in terms of the mechanisms of consultation struck for the negotiation of the FTA and Uruguay Round of the GATT.[25] But the significance of trade concerns has caused governments to broaden representation on consultative forums to all segments of the food chain, most notably food processors. The heightened public attention to the quality and safety of food, and to environmental sustainability, has brought other interests into the agricultural policy community, most notably consumers. In short, the agricultural policy community is now more pluralist and more open, and producers have to defend their concerns before many more potentially competing interests.

Ironically, the farm community is organizationally ill-equipped to take advantage of these consultative opportunities and to articulate a common interest. The farm lobby in English Canada has fragmented in the last fifteen years; the formerly dominant farm federations have lost their leadership status and find themselves competing with commodity groups, which have withdrawn as members and expanded in numbers and policy expertise. Specialist commodity organizations have even shown some capacity to come together under umbrella organizations that compete with the farm federations (the Canadian Federation of Agriculture, for example) for the ear of government. The disunity in the farm community, however, is not consistent. In contrast to the fragmentation in English Canada, the farm lobby in Quebec remains cohesive, as the Union des producteurs agricoles (UPA) enjoys a legal monopoly to represent Quebec farmers' concerns. The strong Quebec farm lobby, with its commitment to an interventionist state, provides an important counterweight to the commodity groups' general preference for a more market-responsive agriculture. But viewed as a whole, with the exception of the province of Quebec, farm groups in the early 1990s appear less capable of accommodating within their ranks the divergent ideological and economic goals across commodity producers, and hence less able to present a united front to governments.

To summarize, since the mid-1980s one can detect four broad changes of import to the agricultural policy-making process: first, the penetration of international trade concerns into domestic agricultural policy; second, new national policy-making structures that have shifted the locus of agricultural policy-making away from Agriculture Canada and opened up the policy community to interests other than producers and provincial governments; third, shifting goals and priorities, which inject the objectives of fiscal restraint, market responsiveness, self-reliance, and favorable relations with trading partners more emphatically into agricultural policy; and fourth, internal cleavages and disunity in the farm lobby.

From one perspective, these domestic and international developments cumulatively appear to suggest a greater autonomy for the national government from producer groups and provincial governments. The negotiation of trade agreements and control over key trade policy instruments (tariffs, import controls, and import licenses) is within federal jurisdiction, suggesting an undermining of provincial influence on agricultural policy. Most provinces, however, are not opposed to the federal market-responsive/trade-liberalization strategy. For producers, the diffusion of policy making away from Agriculture Canada has increased costs for lobbying groups. In addition, the highly technical nature of trade negotiations puts a premium on policy expertise, giving the edge to specialist commodity groups that often enjoy greater financial resources (and hence the resources to hire policy experts) than the farm federations, owing to their government-sanctioned right to collect producer levies for marketing purposes. In any event, the fragmentation in the farm community seems to give an advantage to government, allowing it to pick and choose from among the competing producer-group goals those that are most congruent with its own economic, political, and ideological objectives.

Viewed from another perspective, however, the federal government is still constrained in its freedom and capacity to pursue its own agricultural priorities. This is especially true for its objectives of international trade liberalization and a more market-responsive agriculture. Although the federal cabinet enjoys prerogative powers to negotiate and ratify trade agreements, and so is not legally bound to obtain provincial agreement for international treaties, its ease and effectiveness in subsequently implementing these treaties is very much conditional upon at least prior consultation, if not substantial provincial concurrence with the terms of these treaties. Constitutional and political realities dictate such consultation.

Existing jurisprudence casts doubt on the ability of the federal government to implement provisions in international agreements that impinge upon provincial jurisdiction, at least not without prior consultation and consent of the provinces.[26] Federal success in implementing any international agreement to liberalize trade will thus require provincial cooperation in dismantling the trade-distorting measures (provincial stabilization programs and provincial marketing boards, for example) that their jurisdictional authority over agriculture has allowed provinces to erect.

But constitutional requirements are likely less compelling a reason for provincial consultation than are pragmatic politics; the close links that exist between producers and provinces (in Quebec particularly) mean that a federal government that tried to implement a multilateral agricultural agreement against the wishes of a number of provinces could well find itself with minimal public support in the agricultural community. Recognizing the need to consult and seek the agreement of provinces, Ottawa has augmented provincial input into trade and nontrade policy issues. Provincial ministers of agriculture and their officials are regularly briefed and consulted on the MTN by a subcommittee of the Committee on Multilateral Trade Negotiations (CMTN), which reports to the minister of international trade. The success that the provinces enjoyed in influencing the provisions of the FTA is likely to be replicated in the MTN.[27]

Nor is it easy to dismiss the potential influence of producer groups. Rural ridings are well represented in, and key to the electoral fortunes of, Conservative provincial governments in Alberta, Saskatchewan, and Manitoba. (Approximately half of Saskatchewan's caucus is composed of farmers or individuals with an expertise in agriculture.) With a few exceptions, the agrarian-dominant constituencies in Ontario and Quebec overwhelmingly support the governing Liberal parties.

Farm groups enjoy representation on trade-policy formulation on the Agriculture, Food and Beverage Sectoral Advisory Group on International Trade (SAGIT), a committee that reports to the minister of international trade and which is regularly briefed on the progress of the MTN. This formalized structure is supplemented with more informal briefings by agricultural trade officials in the Multilateral Trade Negotiations Office (MTNO).[28]

Abundant opportunities for consultation characterize the domestic policy review process, entitled "Growing Together," initiated by federal Agriculture Minister Donald Mazankowski in late 1989. Stimulated in part by the government's commitment in the MTN to consider greater access to Canadian markets and removal of export and domestic subsidies, the agriculture minister has struck numerous task forces and committees to recommend reforms in such domestic policies as safety nets, transportation, and farm credit. Amply represented on these task forces, producers will have the opportunity, alongside nonproducers, to make their viewpoints heard.

With both themselves and provincial governments active participants in the consultative processes that surround the making of trade policy and the reform of domestic policy, there seems little likelihood that producers will be frozen out of policy-making. Rather, the system continues to be one of "multiple cracks" of influence.[29] But the question remains: Which policy networks and hence which agricultural programs will be most susceptible to change in this altered political and economic context?

The Durability of Existing Policy Networks and Agricultural Programs

Canadian dairy and egg producers subject to market regulation are most threatened by domestic and international trends. The successful lobby mounted by the dairy sector during the negotiation of the FTA to secure supply management by persuading the national government to add yoghurt and ice cream products to the Import Control List, has come undone in the wake of the GATT Panel Ruling, which has determined that such controls are not legally sanctioned by Article X1.[30] The FTA and the consequent GATT Panel Ruling have highlighted the inherent tension between supply-managed producers and processors of dairy and poultry products, who must compete with lower-cost American imports, produced in the absence of regulated producer prices. The United States is maintaining its pressure on the border-protection feature of supply management in the MTN, where it and other nations are demanding the elimination of such import barriers.

At home, there are sustained pressures for reform of supply-management marketing boards. The national Conservative government's philosophical opposition to market management makes it sensitive to economists' criticisms that marketing boards have inflated quota values that bar the entry of new producers and raise consumer prices. In addition, the government's objective of intergovernmental cooperation makes it anxious to terminate interprovincial discord over quota shares. The budgetary pressures of restraint pose an additional challenge to dairy producers who benefit from a domestic subsidy.

As formidable as these domestic and international threats are, the corporatist policy networks here will probably prove effective in restricting change to some nibbling away and minimal damage. The essential features of supply management are likely to be retained because a number of factor serve to buttress them. The first is a strong producer lobby centered in the national peak associations that represent the dairy and individual poultry sectors, and that has mounted a coordinated campaign during the MTN. The second is the economic significance of these commodities to the provinces of Quebec and Ontario, regions whose electoral support is crucial to the formation of a national government. The third support is the formidable campaign for retention of supply management by the Canadian farm group with the best resources (both in terms of finances and policy expertise), the Quebec UPA and its close alliance with the Quebec provincial government. The final support is the electoral importance of Quebec (and to a lesser degree, Ontario) to the formation of the Mulroney majority governments of both 1984 and 1988, majorities built on solid backing from rural areas.

In the sectors characterized by state assistance—the grains and oilseeds commodities—international and domestic pressures conspire to jeopardize the Crow Benefit subsidies to the railways. Targeted in the Free Trade negotiations by the U.S. as an export subsidy, the WGTA payments to the railways remain highly vulnerable, given that the elimination of export subsidies constitutes a priority for the U.S. in the Uruguay Round of the GATT. Moreover, the domestic interests that opposed the passage of the WGTA in 1983, but that were unable to prevail—cattlemen, the Alberta government, the grain commodity groups—now seem to be in the ascendancy. Bolstering their cause is the government's obvious desire to eliminate the $700 million annual payments to the railways from its budget.

But while international and domestic budgetary and political pressures portend the elimination of freight-rate subsidies, the significance of grains and oilseeds to the prairie regional economies and the political representation of these rural areas in the governing provincial and national caucuses should preserve state-assisted safety nets, albeit safety nets that make less of a call on government coffers. Although the highly organized state of the grain and oilseed sector will ensure that producers have significant input into determining what policies will replace the malfunctioning grain stabilization and crop insurance programs, the prairie lobby's pluralist and contradictory goals will afford the national government considerable latitude to pursue its own objectives regarding new income safety nets.

The trade policies of the U.S. constitute the biggest short-term catalyst to change in the red-meats sector, as rulings by U.S. trade remedy authorities have deemed payments to hog producers under the ASA to be trade-distorting and hence, countervailable.[31] The decision will likely necessitate redesigning this safety net. Largely immune to other multilateral pressures to dismantle export subsidies and import barriers—because of its market-oriented thrust—the greatest challenges to the red-meat sector will come in the medium term from those new entrants to the domestic agricultural policy community, consumers and environmentalists, demanding hormone-free foodstuffs, and animal-friendly and environmentally sustainable production practices.

In conclusion, in a climate where change is, if not imminent, almost certainly unavoidable, producers vary in terms of their ability to steer the direction of change. For their part, international and domestic factors have conspired to give the national government what appears to be an unprecedented degree of autonomy and the capacity to maneuver change in the direction it desires.

Notes

1. A fuller discussion of the following summary can be found in Grace Skogstad, *The Politics of Agricultural Policy-Making in Canada* (Toronto: University of Toronto Press, 1987), chapter 3.

2. This profile of agricultural goals draws upon a number of sources including the following: Vernon C. Fowke, *Canadian Agricultural Policy* (Toronto: University of Toronto Press, 1946); Report of the Federal Task Force on Agriculture, *Canadian Agriculture in the Seventies* (Ottawa: Information Canada, 1970), chapter 3; *Orientation of Canadian Agriculture*. A Task Force Report, II (Ottawa: Agriculture Canada, 1977), pp. 1-57; *A Food Strategy for Canada* (Ottawa: Agriculture Canada and Consumer and Corporate Affairs, 1977).

3. Fowke, pp. 9, 272; *Canadian Agriculture in the Seventies*, pp. 29-31; *Orientation of Canadian Agriculture*, II, pp. 1-9; Figure 1, p. 11.

4. *Canadian Agriculture in the Seventies*, p. 34. Note that the preservation of the efficient family farm is not mentioned as a goal in the 1977 Task Force Report, *Orientation of Canadian Agriculture*.

5. The fruit and vegetable sector also benefited from seasonal tariff protection prior to the implementation of the Free Trade Agreement, but horticultural marketing boards have powers only to negotiate, not set, prices.

6. Economic overviews of national supply management can be found in J.D. Forbes, R.D. Hughes, and T.K. Warley, *Economic Intervention and Regulation in Canadian Agriculture* (Ottawa: Economic Council of Canada and Institute for Research on Public Policy, 1982), and Peter L. Arcus, *Broilers and Eggs* (Ottawa: Economic Council of Canada, 1981). For a discussion of dairy policy, its origins, and functioning, see the following: V. McCormick, "Dairy Price Support in Canada, 1962-1972," *Canadian Farm Economics*, 7, 4 (1972), pp. 2-7; D. Peter Stonehouse, "Government Policies for the Canadian Dairy Industry," *ibid.*, 14, 1-2 (1972), pp. 1-11; and V. McCormick, "Canadian Dairy Policy—The Seventies," *ibid.*, 15, 6 (1980), pp. 1-8. See also Skogstad, *The Politics of Agricultural Policy Making*, chapter 5.

7. *Growing Together* (Ottawa: Agriculture Canada, 1989), p. 55.

8. For a discussion of the organization of processors and producers at the time, see Michael M. Atkinson and William D. Coleman, "Corporatism and Industrial Policy," in Alan Cawson, ed., *Organized Interests and the State* (Beverly Hills: Sage, 1985), pp. 33-4.

9. As quoted in Don Mitchell, *The Politics of Food* (Toronto: James Lorimer, 1975), p. 122. See also Skogstad, *The Politics of Agricultural Policy-Making*, pp. 45-8.

10. A.E. Safarian, *Canadian Federalism and Economic Integration* (Ottawa: Information Canada, 1974), pp. 48-57.

11. Grace Skogstad, "The Farm Products Marketing Agencies Act: A Case Study of Agricultural Policy," *Canadian Public Policy*, 6, 1 (Winter 1980), pp. 89-100.

12. Skogstad, *The Politics of Agricultural Policy-Making*, pp. 90, 97, 117-19.

13. The most unrelenting criticism has come from agricultural economists who endorse a classical economic model. See, among others, Forbes et al., *Economic Intervention and Regulation in Canadian Agriculture*, pp. 100, 113.

14. Historical examinations of grains policy can be found in Vernon C. Fowke, *The National Policy and the Wheat Economy* (Toronto: University of Toronto Press, 1957); Vernon C. Fowke and D.V. Fowke, "Political Economy and the Canadian Wheat Grower," in Norman Ward and Duff Spafford, eds., *Politics in Saskatchewan* (Toronto: Longmans, 1968), pp. 208-20; and Charles Wilson, *A Century of Canadian Grain* (Saskatoon: Western Producer Prairie Books, 1978). Economic discussions of contemporary policies include: Murray Fulton, Ken Rosaasen, and Andrew Schmitz, *Canadian Agricultural Policy and Prairie Agriculture* (Ottawa: Economic Council of Canada, 1989), especially chapter 7; and M. Fulton, "Canadian Agricultural Policy," in Proceedings of the 1986 Annual Meeting, Canadian Agricultural Economics and Farm Management Society, *Canadian Journal of Agricultural Economics*, 34 (May 1987), pp. 109-26.

15. Skogstad, *The Politics of Agricultural Policy-Making*, pp. 40-1.

16. Ibid., pp. 39-40.

17. Ibid., chapter 6.

18. Ibid., chapter 4.

19. For a discussion of corporatism generally and its specific application to the dairy sector, see Atkinson and Coleman, "Corporatism and Industrial Policy," pp. 25-36, and Michael M. Atkinson and William D. Coleman, *The State, Business, and Industrial Change in Canada* (Toronto: University of Toronto Press, 1989), pp. 176-82.

20. Among the several discussions of pressure pluralism, a brief and helpful one is Atkinson and Coleman, *The State, Business and Industrial Change in Canada*, pp. 82-3.

21. Good discussions of the events leading to the Uruguay Round of the GATT and the positions of the major negotiators can be found in: J.C. Gilson, *World Agricultural Changes: Implications for Canada* (Toronto: C.D. Howe Institute, 1989), chapter 6 and pp. 201-5; William M. Miner and Dale E. Hathaway, eds., *World Agricultural Trade: Building a Consensus* (Montreal: Institute for Research on Public Policy, and Washington: Institute for International Economics, 1988), pp. vii-110; T.K. Warley, "Issues Facing Agriculture in the GATT Negotiations," *Canadian Journal of Agricultural Economics*, 35 (1987), pp. 515-34; M.N. Gifford, "A Status Report: The Comprehensive Proposals," Notes for an address to the Conference on "Agriculture and the Uruguay Round of GATT Negotiations: The Final Stages," University of Guelph, Guelph, Ontario, February 20, 1990; and T.K. Warley, "Implications for Canadian Agrifood," Paper for the Conference on "Agriculture and the Uruguay Round of GATT Negotiations: The Final Stages," University of Guelph, February 20, 1990.

22. Fulton et al., *Canadian Agricultural Policy and Prairie Agriculture* document the nature of the crises. See especially chapters 3, 5, and 10.

23. The theme of fiscal restraint itself was not new to the Conservative government. The federal Liberal government had cut back its agricultural spending, which peaked in 1977-78 and thereafter declined to stabilize at 1974-75 levels. See D. Berthelet, "Agriculture Canada policy and expenditure patterns, 1868-1983," *Canadian Farm Economics*, 19, 1 (1985), p. 12 and graph 4, p. 13. For a discussion of the altered

perspectives of the Conservative government, see Grace Skogstad, "Agriculture: Sharing the Responsibility," in Michael J. Prince, *How Ottawa Spends 1987-88: Restraining the State* (Toronto: Methuen, 1987), pp. 268-92. The clearest and most recent statement of the Conservative government's emphases on self-reliance and market responsiveness is *Growing Together* (Ottawa: Agriculture Canada, 1989).

24. J. Stefan Dupré discusses these developments in "Reflections on the Workability of Executive Federalism," in Richard Simeon, Research Coordinator, *Intergovernmental Relations*, Volume 63, Royal Commission on the Economic Union and Development Prospects for Canada (Toronto: University of Toronto Press, 1985), pp. 1-32.

25. These institutions are described by Gilbert Winham, "Formulating Trade Policy in Canada and the US: The Institutional Framework," in Grace Skogstad and Andrew Fenton Cooper, eds., *Agricultural Trade Policy: Domestic Politics and International Tensions* (Montreal: Institute for Research on Public Policy, 1990).

26. See Anonymous, "Issues of Constitutional Jurisdiction," in Peter Leslie and Ronald Watts, eds., *Canada: The State of the Federation 1987-88* (Kingston: Institute of Intergovernmental Relations, 1988), pp. 45-6; and the testimony of Law Professor Andrew Petter to the Senate Standing Committee on Foreign Affairs, *Proceedings*, February 23, 1988, p. 15:7, and p. 15:42 on the "extent of obligations clause" in GATT (Article XXIV:12).

27. Douglas M. Brown, "The Federal-Provincial Consultation Process," in Leslie and Watts, *op. cit.*, pp. 77-93, documents provincial involvement. The conclusion regarding provincial influence relies upon interviews with individuals negotiating the agricultural provisions in the FTA.

28. For a fuller discussion, see Grace Skogstad, "Canada: Conflicting Domestic Interests in the MTN," in Skogstad and Cooper, *Agricultural Trade Policy*.

29. The "frozen out" thesis suggests that once federal and provincial governments begin bargaining (usually behind closed doors), interest groups are shut out and their goals often ignored as government interests take priority. The "multiple crack" thesis states that federalism, with the provision of powerful governments at two levels, offers interest groups two opportunities to prevail, at both the provincial and federal levels.

30. General Agreement on Tariffs and Trade, *Canada-Import Restrictions on Ice Cream and Yoghurt*, Report of the Panel, September 27, 1989.

31. *Live Swine and Pork from Canada*, Determination of the Commission in Investigation No. 701-TA-224 (Final . . . US ITS Pub. 1733, July 1985, A-21, A-22).

3

The Political Economy of Agriculture in the United States

Gordon C. Rausser

U.S. agricultural programs have resulted in enormous budgetary costs, huge surpluses of farm products, major trade disputes with other countries, great harm to well-functioning international markets, and benefits that do not reach those most in need. Between 1981 and 1985, the federal government spent about $60 billion on farm price- and income-support programs. In 1986 alone, the actual cost of commodity credit corporation activities to support the agricultural sector was $25.9 billion. Other programs also support agricultural production in rural America; in 1986, the outlays for these programs amounted to approximately $14 billion, the largest single item being the farmer home administration outlay at $7.3 billion. Again in 1986, the value of commodities exported under Public Law (PL) 480 (both Titles I and II) amounted to $1.4 billion, export guarantees came to $2.5 billion, and the export enhancement program that subsidizes U.S. agricultural sales abroad cost the U.S. government nearly $0.75 billion. In the domestic market, food assistance programs cost $20.2 billion in fiscal year 1986; the bulk of this expenditure supports the Food Stamp Program and the Women, Infants and Children Feeding Program. Finally, as a result of special legislation in December 1987, the U.S. government assumed a contingent liability for the Farm Credit System (composed of the Federal Loan Bank, Production Credit Association, and the Bank for Cooperatives) of $4 billion.

Are these massive governmental interventions the result of productive policies that correct for market imperfections, lower transaction costs, or effectively regulate externalities? In other words, is the U.S. government acting as the benign, perfect instrument that is presumed in conventional welfare economics? Or, are these programs the result of manipulation by powerful commodity or agricultural interest groups actively engaged in rent seeking or directly unproductive activities (Buchanan and Tullock, 1962; Krueger, 1974; Bhagwati, 1982)? In contrast to the tradition of Pigou and Mead, this perspective does not regard the state or public sector as a benevolent guardian of the public interest. Machiavelli and Hobbes are its inspiration. In effect, agricultural interest groups are presumed to behave much like the proverbial 800-pound gorilla—he walks where he wants, he stands where he wants, he sits where he wants, and he gets what he wants. In the case of U.S. agricultural policy, it will be argued that these two extremes only set the outside bounds on actual government behavior. Accordingly, a resounding no must be the response to both of the questions posed.

Agricultural policy at the level of the federal government is a complex web of interventions covering output markets, input markets, trade, public good investments, renewable and exhaustible natural resources, regulation of externalities, education, and the marketing and distribution of food products. Even a superficial analysis of this wide cadre of policies demonstrates that neither of the two extremes found in the economic literature provides either a positive or, for that matter, a prescriptive framework. Instead, such an assessment leads to the view that, in the case of agriculture, the public sector has been both "productive" and "predatory." A formulation in the literature that treats both types of behavior in an internally consistent framework is the PESTs (predatory) and PERTs (productive) formulation. PEST policies, or political economic-seeking transfers, are meant to redistribute wealth from one social group to another and are not explicitly concerned with efficiency. In contrast, PERTs, or political economic resource transactions, are intended to reduce transaction costs in the private sector by, for example, correcting market failures or providing public goods; they are ostensibly neutral with respect to distributional effects.

A historical review of public policy in agriculture reveals not only tension between the PERT and PEST roles of the public sector, but also some coordination between these two types of activities. This historical sketch is presented under the heading "History," to motivate the joint determination perspective of both PERT and PEST policies, which are discussed later in the chapter. The link between these two types of policies is made in a world of rational decision makers, where actual policy selection reflects an optimization game that can be modeled as the maximization of a "governing criterion function." Policies are in place, in part because they serve the interests of those with relative political power and influence. The government, however, has some autonomy and cares not only about the distribution of the pie but its size as well. The rational process generates a portfolio or mixture of productive (PERT) and predatory (PEST) policies. There is a wide scope of possibilities to interchange the use of PESTs and PERTs so as to acquire, balance, and secure political power. The argument will be made that the PERT/PEST integrated framework provides the foundation for operational and meaningful prescription.

History

Productive (PERT) Policies

Over much of the last century and a half, the agricultural sector has been one of the most innovative and productive sectors of the U.S. economy. During the early part of this period, the public sector played a major role in reducing transaction and information costs. This helped lead to a dramatic increase in the size of markets, and specialization among individual producers and regions. The period from 1850 through 1880 witnessed the emergence of a number of important institutions formed with the intent of lowering transaction costs in the private sector (Morrill Act, 1862; U.S. Department of Agriculture, 1862; the Hatch Act, 1887; and the Smith-Lever Act, 1914). At the end of the century and during the early 1900s, farm interests sought fundamental changes in the rules of the game and in the use of federal power for distributing wealth and income in their favor. These desires were expressed in various forms, viz., easy money created by governmental action, government funds supplied for farm mortgages, and government guarantees of farm commodity prices.

There can be little doubt that, during the 1920s and 1930s, farmers became one of the most well-organized economic interest groups. The national system of county agents, the American Farm Bureau, and the U.S. Department of Agriculture (USDA) all combined with a clarity and singleness of purpose to promote economic growth for the U.S. agricultural sector. The grassroots organization represented by the county agent extension system proved to be an effective vehicle for systematically communicating new agricultural technologies and knowledge directly to farmers. It also became the vehicle for communicating information on farmers' problems requiring research back to the USDA and the land-grant universities. During this period, farm interests were also effective in evading a number of governmental interventions, with exemptions from antitrust, labor, and tax legislation.

A significant portion of the increased growth of the U.S. agricultural sector over the last century can be traced to the provision of public goods and governmental investments in infrastructure. State and federal support of land-grant universities has certainly had a positive influence on the level and quality of human capital. Transportation system investments, water resource developments, and land reclamation activities have all made significant contributions to economic growth. Pavelis (1985) has estimated that, from 1900 through 1975, "the federal government, through direct construction or indirect cost sharing, has created up to 1975 about 45 percent of the value of all irrigation, drainage, and soil and water conservation facilities in the United States."

A major resource available to the U.S. agricultural sector has been the knowledge generated by public-sector investments in research, which has been shown time and again to have a significant influence on agricultural productivity and growth. The striking feature of agricultural research policies has, however, been the overwhelming evidence of underinvestment. As Ruttan (1982) has shown, the rates of return to public good investments in agricultural research justify much higher levels of public research support.

Productive Policies with Concentrated Benefits

Early federal government policy can largely be characterized as long-run institutional development. Here, the federal government was supplying public goods whose associated benefits and costs were widely dispersed. From the late eighteenth century through the 1920s, the federal government continued to design and implement growth-promoting policies, which initially improved efficiency in the private sector, but whose benefits were highly concentrated and whose costs were widely shared. This included those regulatory policies that focused on limited information and the unacceptable levels of transaction costs facing some private markets (measure and grade standards, truth in labeling, etc.).

To be sure, however, most agriculturally related PERT policies do more than simply promote economic growth. The view of public work projects is often justified as turning "wastelands" into valuable agricultural soils. The benefits of such public-good investments can be concentrated; as a result, the public sector has on occasion attempted to limit the concentration of redistributed wealth or income. For example, federal and state policies influencing western water development have been instituted in conjunction with below-cost pricing of irrigation water. Because of the potential concentration of transfers resulting from this resource policy, the original federal Reclamation Act of 1902 limited the size of the farms that could receive low-cost irrigation infrastructure and water to 160 acres. The intent of the original legislation was to ensure that water projects benefited smaller farms rather than powerful land interests such as railroads, oil companies, or land speculators. Initially, these water resource projects were designed to be self-financed, and project beneficiaries were expected to repay construction costs over a ten-year period (Englebert and Scheuring, 1982; Holmes, 1979; Worster, 1985).

Almost from the outset, effective political influence was exercised to alter the original provisions of the 1902 legislation. As a result of one financial crisis following another, the self-financing features were eliminated and, by 1930, reclamation construction funds were appropriated from the general treasury. Moreover, during the 1930s, through a series of congressional enactments, the scope of reclamation policy was greatly increased to include activities such as fish and wildlife habitat development, flood control, navigation, and hydroelectric power generation and distribution. By the 1940s, the original intent (promoting small-scale farmer settlement of the arid West) of the 1902 legislation was masked. By the end of the 1940s, the "Iron Triangle," consisting of the U.S. Bureau of Reclamation, which served large-scale farming interests who, in turn, supported development-minded congressional representatives, was firmly entrenched.

In the early part of the century, in addition to reclamation and water development, Congress passed a number of other bills motivated, in part, by the desire to enhance efficiency and lower transaction costs. This legislation covered rural delivery of mail, soil conservation, agricultural credit, rural electrification, rural road building, and many other investments in the physical infrastructure and inputs of agriculture. These growth-promoting policies generated benefits that were not available to all citizens or even to all farms. There is little doubt that some laws were administered to create selective benefits for specific groups. Over the same period, other legislation regulating different aspects of agriculture—e.g., fertilizer and seed standards, weights, animal health, and food safety—was introduced, again, to lower other types of transaction costs. These regulations imposed direct cost on some and major benefits on other specific groups in society. A few examples—farm credit, environmental pesticide, and soil conservation policies—are examined in the following paragraphs.

Farm credit legislation (federal land bank, production credit association) has been justified in some circles by different types of market failures. One failure, government based, results from banking laws that have prohibited the emergence of a national bank. State banks cannot easily diversify the significant regional components in agricultural lending risk. A second alleged market failure is the high cost of assessing farm credit; a third is the principal-agency problem that can emerge in related equity markets. A fourth is the adverse selection that may exist in rural credit markets due to farmers having heterogeneous abilities that are unknown to bankers or investors. To the extent that the original legislation addressed these potential problems, transaction costs were reduced in rural credit markets and economic growth was enhanced.

The institutions that were established by this legislation, particularly the system of federal land banks and production credit associations, became government-sponsored enterprises. The design of the farm credit institutions failed to place equity capital at risk, and the residual claimant features were poorly defined. The stockholders of the federal land and production association banks were also its borrowers, creating an obvious conflict of interest. Since stock could not be freely traded and could only be redeemed by paying off loans, another adverse selection problem arose when the entire system faced a financial crisis in the 1980s. The crisis was more severe than it would otherwise have been because of the enabling legislation that restricted the system's activities to the agricultural sector, rendering any other source of risk diversification unavailable. The system also put in place what proved to be unworkable liability features.

Even though the original legislation may have been motivated by market failure concerns of one type or another, as the years unfolded, the farm credit system invested in and established a lobbying organization that represented the narrow interests of the farm credit system. In 1985 and 1986, when the system lost an average of over $2 billion per year, millions were spent per year to finance lobbying efforts. These expenditures, and the effectiveness of the lobbying organization, are partially responsible for the federal government bailout of the farm credit system in 1988.

The origins of environmental pesticide policy begin with the Federal Insecticide Act of 1910, developed to protect farmers from fraudulent claims of insecticide salesmen (Bosso, 1987). In essence, this legislation was introduced to lower the transaction costs resulting from imperfect information. Today, the environmental and health hazard regulatory portfolio includes surface water pollution; groundwater pollution; air pollution; worker exposure to agricultural chemical inputs; endangered species (exposure to the harmful effects of pesticides applied to the fields and crops in their habitat); and dietary risk (pesticide residues may remain in agricultural products that reach the consumer).

As with water reclamation policy, pesticide environmental policy of the 1950s and 1960s was firmly controlled by the Iron Triangle comprised of the pesticide industry, the USDA, and Congress (Bosso, 1987; Mitchell, 1979; Macintyre, 1987). As public awareness began to increase, this Iron Triangle was challenged first by the Pesticide Control Amendment of 1954 that required any registered pesticide to have a tolerance level for acceptable residues set by the U.S. Food and Drug Administration (FDA). A far more important event, however, was the passage in 1958 of the Delaney Amendment, which states simply that "no (food) additive shall be deemed safe if it is found to induce cancer when ingested by man or animal." As expected, this amendment was vigorously opposed, without success, by agricultural chemical interests, and passed easily.

Until the publication of Carson's (1962) *Silent Spring*, most people knew the benefits of pesticides, but very few had any knowledge of the possible environmental and health risks of pesticide use (Perkins, 1982). The Carson message received much credibility with the thalidomide scare of 1962. Slowly, a new breed of environmental activists emerged, which turned to the court system to enforce the laws that had been enacted. In 1969, the Environmental Defense Fund won a case against the use of DDT in the state of Wisconsin. As Bosso (1987) reported, this first state-level ban of DDT "sent shock waves throughout the community, the chemical industry, and government at all levels."

With the Nixon administration's 1969 announcement of its intent to phase out all nonessential uses of DDT within two years, the Iron Triangle, weakened in the 1960s, was shattered, quickly losing influence to environmental interest groups armed with a number of significant events, anecdotal evidence, and a responsive court system. These interest groups have expanded their portfolios to include the external effects of scientific agriculture on the nation's waters and wildlife habitat, in addition to the quality of food supply. They have helped usher in modern pesticide policies whose foundation includes the National Environmental Policy Act of 1969, the Clean Air Act of 1970, and the establishment in 1970 of the U.S. Environmental Protection Agency (EPA) within the Executive Branch. In terms of the political economy, a new triangle has emerged consisting of the environmental interest group organizations, the EPA, and members and committees of Congress without agricultural ties.

The Masking of Predatory with Productive Policies

When the Supreme Court declared the Agricultural Adjustment Act of 1933 unconstitutional, the first overt attempt to join input and output market policies emerged with the Soil Conservation and Domestic Allotment Act of

1936. The latter dropped the processor tax financing provision of the 1933 act and used soil conservation as an instrument of commodity supply management. Specifically, this act enabled farmers to receive soil conservation payments for reducing "soil depleting" crops which, unsurprisingly, were also surplus crops. The partial focus on conservation emanated from concern about preserving and sustaining agricultural lands for future generations, "soil mining" and erosion, and "soil runoff" externalities. Although the implementation of commodity supply management and soil conservation policy was temporarily annulled by the 1938 Agricultural Adjustment Act, 1936 marks the beginning of a long history of resource enhancing and commodity redistribution policies—at times simultaneously determined, at other times sequentially determined, but always highly interactive.

The soil bank, established by the Agricultural Act of 1956, was the next major effort toward commodity redistributive and soil conservation policies. This legislation attempted to bring about adjustments between supply and demand for agricultural products by taking farmland out of production. The program was divided into two parts—an acreage reserve and a conservation reserve. The objective of the former was to reduce the amount of land planted to allotment crops: wheat, cotton, corn, tobacco, peanuts, and rice. In contrast to the acreage reserve, all farmers were eligible to participate in the conservation reserve. This long-term general retirement program allowed conversions of whole farms from cropland to soil-conservation areas. This program eventually enrolled nearly 30 million acres in the 1960s, moving marginal cropland into permanent pasture, timber, or recreational uses under contracts for a maximum of ten years; it served the objectives of encouraging long-term adjustment of land and labor to nonfarm uses, soil conservation, and to some degree, output management. Most of the land in this conservation reserve, however, returned to production during the 1973-1975 boom.

In 1985, conditions were once again ripe for a combination of commodity redistribution and soil conservation resource policies. The exorbitant cost of commodity programs, the unanticipated economic events that occurred throughout the early 1980s—huge program commodity surpluses and an increasingly more effective environmental lobby—all combined to lead to the establishment of a new conservation reserve program under the 1985 Food Security Act. The 1985 Act succeeded in reducing acreage bases of corn, wheat, and cotton by 1.1, 2.5, and 8.2 percent, respectively, from 1985/86 to 1987/88 (Tweeten, 1989).

A number of important lessons, only four of which are outlined here, have emerged from the evolution of commodity and soil conservation policies since the 1930s. First, political support can be generated for redistributive commodity policies (in the 1933 Agricultural Adjustment Act) when they are masked by public interest policies that protect future generations and promote environmental quality (e.g., the Soil Conservation and Domestic Allotment Act of 1936, the Agricultural Act of 1956, and the Food Security Act of 1985). Second, the combining of commodity and resource policies would have been far more difficult if institutional investments had not been undertaken to establish soil conservation districts and/or the county agricultural adjustment committees (which, in 1981, were relabeled the agricultural stabilization and conservation committees). These institutional arrangements provided an effective organization through which local farmers and the federal government could join forces to implement soil conservation practices. These practices have resulted in terracing, strip cropping, drainage crop rotation, contouring, fertilization, pasture improvement, control grazing, tree plantings, and so on.

Third, to promote public interest conservation practices, farmers have had to be compensated for production restraint, but these compensation schemes have only been acceptable during times of depressed markets. In rapidly expanding markets, public compensation for conservation practices has not generated sufficient political support. This, in part, is why commodity policies have been coupled with soil conservation policies during periods of depressed market conditions, but not when they are favorable.

Fourth, in its role of resolving conflicts among alternative economic interest groups, governments often design programs that appear incoherent. In one program, for example, conservation is promoted (which often requires the retirement of vulnerable acreage), while another program offers price supports based on historical acreage (a system that penalizes premature land retirement). These apparent contradictions, however, are often the direct result of effective institutional arrangements which, in turn, generate sufficient support for governmental action.

Predatory (PEST) Policies

With the passage of the Agricultural Adjustment Acts of 1933 and 1938, the long history of redistributive agricultural policies began in explosive fashion. These policies implemented "coupled" transfer schemes that directly benefited concentrated interest groups. They were nonneutral with respect to production and, as a result, farmers could influence the amount of wealth and/or income transfers through their actions.

With the introduction of these programs in the 1930s, the generic farmer organizations began to lose influence and, with the growing specialization across commodity lines, economic interest groups became more concentrated. The legislation of the 1930s also created direct economic benefits or losses for particular groups and invited these groups to become actively involved in setting specific levels of commodity policy instruments. They, in effect, became the primary vehicle for political expression of farmer interests (Lowi, 1965). Similarly, beginning in the 1930s and through the postwar years, the USDA was transformed from an organization that focused largely on research and education to a more conventional government agency managing programs that provided direct economic benefits to specific interests.

With the commodity redistributive policies come, of course, deadweight losses and contractions in economic growth. The range for these losses has been estimated for many commodities on numerous occasions. Neglecting the waste generated from rent-seeking behavior, an indication of the range of these deadweight loss estimates for selected commodities is reported in Table 3.1. The stylized facts emerging from such analyses (not only for the

Table 3.1. Annual Gains and Losses from Income-Support Programs Under the 1985 Food Security Act and Trade Restrictions

Commodity	Consumer Loss	Taxpayer Cost[a]	Producer Gain	Net Loss
	billions of dollars			
Corn	0.5 – 1.1	10.5	10.4 – 10.9	0.6 – 0.7
Sugar[b]				
Case I	1.8 – 2.5	0	1.5 – 1.7	0.3 – 0.7
Case II	1.1 – 1.8	0	1.0 – 1.4	0.1 – 0.4
Milk	1.6 – 3.1	1.0	1.5 – 2.4	1.1 – 1.7
Cotton	c	2.1	1.2 – 1.6	0.5 – 0.9
Wheat	0.1 – 0.3	4.7	3.3 – 3.6	1.4 – 1.5
Rice	0.02 – 0.06	1.1	0.8 – 1.1	0.06 – 0.32
Peanuts	0.2 – 0.4	0	0.15 – 0.40	0.0 – 0.05

[a] Includes CCC expenses after cost recovery.
[b] Case I assumes U.S. policies do not affect world sugar prices. Case II takes into account the fact that U.S. policies reduce world sugar prices. The value of sugar import restrictions to those exporters who have access to the U.S. market (that is, value of quota rents) is $250 million.
[c] Less than $50 million.
Note: All figures reflect Gramm-Rudman-Hollings.

United States but for other countries as well) can be briefly summarized as follows: the redistribution of income to agriculture is greater, the richer or the more industrialized the country; the higher the cost of production; the fewer the number of farmers, absolutely and relative to the total population; the more price inelastic the supply or demand function; the lower the portion of total consumer budget spent on food; and the "smaller" the exporting country or the "larger" the importing country. In general, the accumulated evidence is that the commodity specific policies instituted in the United States and other developed countries involve significant distortions.

Initially, the major policy instruments for achieving redistribution were price supports and public storage. Since price supports were generally set well above market equilibrium prices, accumulation of public storage became a necessary by-product.[1] Because of the large increases in productivity resulting in part from the public investment in research, huge surpluses naturally began to mount.[2]

Along the historical record from the 1930s, mounting surpluses have led to a piecemeal proliferation of policy instruments that address the unintended side effects of excess production. On the supply side, this proliferation includes *inter alia*, land controls, land conservation, production quotas, commodity specific (1983) and later commodity generic certifications (1986)—a device for releasing stocks held in public storage; on the demand side it includes export subsidies, export enhancements, PL 480 (concessional sales and food grants), and the USDA Food Stamp Program. With each additional policy instrument, however, came other unanticipated side effects that required additional vehicles for containing the expanding capacity for commodity production.

Forms of Substitution and "Side Effects"

Many forms of substitution exist both within and across commodities, the effects of which seem to continually surprise both policymakers and interest groups. Accordingly, the experience of coupled PEST policies in U.S. agriculture has shown again and again the impossibility for any particular policy instrument mix to cover all "margins" subject to optimization. The long-run, unintended consequences seem to arise because (1) the "collective" policy-making process is generally short sighted; (2) interest groups involved in PEST-related activities are seldom concerned with the larger external consequences of their actions; and (3) policymakers and interest groups cannot adequately anticipate the full set of possible responses due to the complexity of substitution possibilities.

The current mix of coupled PEST policy instruments is shaped by the unanticipated consequences of numerous incremental policy choices, which continually modify the prior constraints. So long as coupled PEST incentives are sufficiently large, surpluses will be generated. As a consequence, as noted above, there have been many attempts to manage supply by various methods. In terms of substitution possibilities, the initial voluntary acreage reduction programs focused on compliance requirements for a particular commodity, neglecting the supply or, for that matter, demand substitutability with other commodities (Brandow, 1977). This, in turn, led to cross-compliance requirements for related commodities (e.g., corn, soybeans, oats, and sorghum).

Of course, if the direct- and cross-compliance requirements are excessively burdensome, farmers will choose not to participate. In essence, the government attempts to manage total market supply by inducing farmers to participate voluntarily with offers of subsidies, primarily in the form of deficiency payments, and also allowing participants to be eligible for price supports and disaster payments (in counties where federal crop insurance is not available). In return, the government asks farmers to idle some portion of their controlled land. By idling land, the government hopes to reduce the supply to markets, thereby raising prices and indirectly lowering the amount of the deficiency payments and its treasury exposure.

Income transfers to farmers through deficiency payments have nonneutral effects on production through four potentially different channels: target prices, price supports, the land resource base, and productivity. Mechanically, the deficiency payment rate is computed as the difference between the target price (set by law) and the higher of the basic loan rate or the average market price received over the first five months of the marketing year. The payment base is determined by the land base and "program yield" (based on the individual's or counties' past yields) adjusted for any acreage production programs. Specifically, for those who participate, the expected deficiency payment for a particular crop, c,

$$E(d_{ct}) = [P^T_{ct} - \text{Max}(P^S_{ct}, E(P_{ct}))](1 - \omega_{ct})L_{ct}Y_{ct}$$

where P^T_{ct} is the target price; P^S_{ct}, the support price; $E(P_{ct})$, the expected average price received by farmers; ω_{ct}, the percentage of land base required to be idled; L_{ct}, the land base in period t; and Y_{ct} the program yield per unit of the land base. Target prices are set well above market prices which, of course, encourages high program participation. To receive the deficiency payments, participating farmers must allocate land to the program crop (c) or conserving uses dictated by USDA. Moreover, current law generally

The Political Economy of Agriculture in the United States

requires farmers to forego present and future program benefits if they harvest crops other than the program crop for which they have a "land base." This feature, of course, couples the transfers to the planting and harvesting of program crops.

Each of the four components of this transfer scheme $(P_{ct}^T, P_{ct}^S, L_{ct}, Y_{ct})$ provides incentives for augmenting output on the utilized land base. Until the 1986 market year, Y_{ct} was determined as a moving average of a farmer's past yields. One provision of the 1985 legislation, however, was to assign unalterable program yields to Y_{ct}; thus, this potential source of nonneutrality was eliminated. Moreover, the degree of nonneutrality derived from price supports was also dramatically reduced with the 1985 legislation.[3] In the case of land base, however, current actions can influence the computation of L_{ct}. Operationally, a producer of a program crop has an assigned "base" acreage of that crop, which is derived from a five-year moving average of plantings of that crop on the farm. As a result, it is often advantageous for a producer not to participate in a program in order to increase his computed land base in anticipation of higher future subsidies (Foster, 1987). Available evidence reveals that producers with a low land base generally choose not to participate, whereas those with a high land base do.[4]

That portion of the actual land base that is idled is an unconstrained choice of each producer. Since land is not of uniform quality, each producer will rationally idle the least productive land that he controls (whether as an owner or a renter). This factor, plus the existence of nonparticipants and the failure to comply, means that a given percent acreage-reduction program normally reduces output by a substantially smaller percentage. This phenomenon has been referred to as "slippage," a number of different estimates of which exists in the literature, ranging from 30 percent to as high as 60 percent (Norton, 1986; Tweeten, 1989; Love and Foster, 1990; Ericksen, 1986).

Given the structure of the deficiency payments, it should be clear that the treasury exposure for financing commitments is highly uncertain. Since P_{ct}^T is set by the legislation, the major sources of uncertainty are P_{ct}^S, L_{ct}, and the participation rate. Prior to 1986, the level of Y_{ct} used in the computation of deficiency payments was also uncertain over the planning horizon following each four-year (or so) revision in the basic Agricultural Adjustment Acts of the 1930s. Because of this uncertainty, it is not surprising that the Office of Management and Budget and the USDA frequently generate point forecasts for government expenditures that are widely off the mark. What is perhaps more startling, however, is that compared to other budgetary predictions, agricultural expenditures have been forecast with very high errors and have been systematically downward biased over the last few decades. To be sure, there are incentives for underestimating the expected treasury costs in those areas where the exposure is highly variable and the PEST recipients are distinctly more powerful and better informed than those who share the burden for the PEST transfers (i.e., taxpayers, consumers).

Because of equity concerns and the desire to control the treasury exposure, deficiency payments are limited to $50,000 per farm, while loan deficiency payments (based on the difference between the basic loan rate and the secretary of agriculture's announced loan rate) were limited to $200,000 for the 1987 crop. Over the years, however, there have been many loopholes that allow these payment limitations to be exceeded. In essence, the loose definition of a "person" has fostered a proliferation of overlapping partnerships and other farm reconstitutions in order to qualify for multiple payment limits. Accordingly, the number of "farmers" in program-eligible commodities has increased over the last decade, a trend which contradicts that found in the noneligible commodity-systems. As a result, the distribution of program benefits continues to be viewed by many as inequitable. As shown in Figure 3.1, government payments are concentrated among the largest farming operations, with the average payment to all farmers having annual sales exceeding $500,000 per year being well above $40,000. Since many large farms do not produce commodities eligible for government programs (approximately 25 percent of all farmland in the United States is eligible for government crop programs), participating farms receive considerably more than this figure.

Over the years, the nonneutrality of the transfer schemes has led to excess utilization of basic inputs. Unfortunately, many of the inputs used in the production of agricultural commodities are joint, producing valuable as well as undesirable outputs (Rausser and Lapan, 1979). The process of agricultural production itself results in the generation of material waste. Residuals of fertilizer and pesticide applications combine with excess water and are transported into various water sources used in the production process. Toxic salts continue to accumulate in agricultural land. Burning crop residues may result in air pollution. The excess utilization and cultivation of land leads to excessive soil erosion. While much of the empirical evidence on these effects are local and depend on specific conditions, national estimates have been made for soil erosion rates experienced on different types of farms. Reichelderfer (1985), for example,

Figure 3.1. Average Direct Government Payments per Farm by Sales Class, 1988

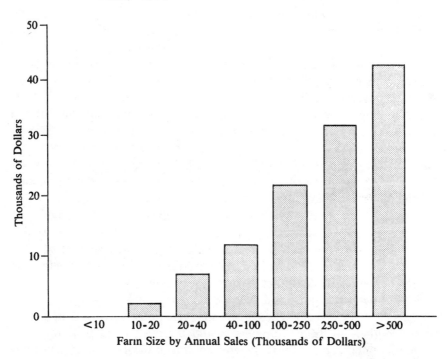

Source: U.S. Department of Agriculture

has estimated an erosion rate of 9.2 tons per acre per year as the average for nonparticipating (in government programs) farms, while all farms that do participate and are recipients of nonneutral transfers have an average erosion rate of 14.7 tons per acre per year. This difference is highly significant.

The above examples illustrate the joint input feature of land, water, pesticides, etc., leading to multiple products—some subset of which results in the degradation of environmental resources. The external cost associated with this process has led to the emergence of environmental policies and various pollution abatement activities within the agricultural sector. Due to the complex problem of monitoring and estimating environmental harm, Pigouvian taxes based on marginal damage are impractical (Rausser and Howitt, 1975). As a result, both measurement or information policies must be put in place with environmental standards, taxes, or control policies (Hochman and Zilberman, 1979).

The inflexible settings of price supports and target prices in the early 1980s followed the favorable economic decade of the 1970s. The coupled PESTs policies augmented the degree of overexpansion within the U.S. agricultural sector, making it especially vulnerable to the unanticipated interest rate, exchange rate, and growth rate scenarios of the early 1980s. Associated with these unintended side effects of inflexible transfer schemes is the "mining of the soil" that many farmers engaged in during the 1980s in order to survive financial stress (Foster and Rausser, 1990). Similar activity is undertaken in order to generate the production history that will enhance a farmer's position with future PEST transfers. This phenomenon has occurred on other occasions and dramatically increases the production of environmental "bads."

Partially because of the nature of the coupled transfers, one production record after another was broken during normal weather years of the 1980s. In 1983, commodity-specific certificates were offered in lieu of cash transfers and became a means for releasing public stocks held in the farmer-owned reserve. This farmer-owned reserve (established by the 1977 legislation) was designed as a means of keeping public stocks off private markets until prices rose sufficiently to motivate farmers to sell their stocks, or still higher, so the government would force their release. In 1986, generic commodity certificates were introduced in place of commodity-specific certificates; this, in effect (up to the level of available stocks), allowed the USDA to "print money."

In the world markets of many commodities the United States is a "large country." Under certain conditions, its accumulation of stocks can result in short-run favorable consequences for all exporters of the commodity in question. In other instances, where the internal price supports are so high as to effectively eliminate the export market as a relevant alternative to U.S. production (without export subsidies), all the benefits accrue to other exporting countries in the short run.

Over much of the post-World War II period, the United States has behaved as a residual supplier on world markets of many major commodities, especially the food grains, cotton, and the feed grains. A number of empirical estimates have been made as to the worldwide price effects of liberalizing agricultural policies in the United States (Tyers and Anderson, 1986; Roningen and Dixit, 1989; and Zietz and Valdez, 1986). For example, Roningen and Dixit estimate that an elimination of agricultural policies in the United States would increase world dairy product prices by 23.5 percent, sugar prices by 22.8 percent, coarse grain prices by 11.6

percent, wheat by 10.6 percent, rice by 2.9 percent, ruminant meats by 3.8 percent, and nonruminant meats by 3 percent.

Mix of PESTs and PERTs

The degree of governmental intervention across commodity groups can be represented by producer subsidy equivalents (PSEs), measured as the ratio of the total value of public sector assistance (whether in the form of PERTs or PESTs) to total receipts. As shown in Table 3.2, PSE levels are indeed dramatic for the sectors with inelastic demands such as sugar, milk, rice, and wheat. Feed grains have an intermediate level of support while sectors with more elastic demands, such as soybeans and red meats, have the lowest level.

The decomposition of the PSEs between PERTs and PESTs, or productive and predatory policies, are also reported in Table 3.2. For the case of PERT investments in public research and PESTs-coupled subsidy transfers, de Gorter, Nielson, and Rausser (1990) show that the joint determination of these two types of policies turns on the efficiency of the transfer (the amount of deadweight loss associated with the coupled PEST policies) and a few key parameters, viz., the productivity of research, the supply and demand parameters for the commodity in question, and the relative power reflected by preference weights across various interest groups. If a PERT policy harms producers because of highly inelastic demand, responsive supply, and high productivity, but producers have more political clout than other interest groups, the amount of public investment in research will be inadequate. The evidence on underinvestment in agricultural research is overwhelming (Ruttan). However, for these noted facts, their framework generates the hypothesis that complementarity exists between PESTs and PERTs policies which, in turn, means that less underinvestment in public research occurs than would otherwise be the case. In essence, subsidies compensate some farmers for potential losses emanating from research expenditures, and thus political obstruction can be effectively countered.

As a percentage of total PSEs, the inelastic demands sectors (e.g., sugar, milk, and rice) have the lowest relative level of PERT support, while the elastic sectors (e.g., soybeans and components of meat complex) have the highest. The recorded data substantiates the view that coupled PESTs policies are higher in sectors with inelastic demand and highly productive research, and lower in sectors with highly elastic demand and low supply

Table 3.2. Productive Versus Predatory Policy Interventions in U.S. Agriculture, 1982-1986 Average

	Total Producer Subsidy Equivalents		Policy Interventions			
			Productive		Predatory	
	Percentage Unit Value	Dollars per ton	Dollars per ton	Percentage Producer Subsidy Equivalent	Dollars per ton	Percentage Producer Subsidy Equivalent
Sugar	77.4	221.0	17.5	7.9	203.5	92.1
Milk	53.9	152.0	11.8	7.8	140.2	92.2
Rice	45.0	145.0	9.4	6.4	135.6	93.6
Wheat	36.5	57.0	7.7	13.5	49.3	86.5
Sorghum	31.5	36.4	5.3	14.5	31.1	85.5
Barley	28.8	32.0	6.7	20.9	25.3	79.1
Corn	27.1	31.5	5.6	17.7	25.9	82.3
Oats	7.6	7.3	4.5	61.6	2.8	38.4
Soybeans	8.5	18.0	13.4	74.3	4.6	25.7
Beef	8.7	175.3	97.7	55.5	78.1	44.5
Poultry	8.3	82.0	53.3	65.0	28.7	35.0
Pork	5.8	85.4	67.6	82.5	17.8	17.6
Average	24.6	86.9	25.1	35.6	61.8	64.4

Source: U.S. Department of Agriculture, "Estimates of Producer and consumer Equivalents: Government Intervention in Agriculture," Economic Research Service, ATAD Staff Report No. AGES 880127, April 1988.

elasticities (e.g., meat complex). In very highly elastic sectors (e.g.,speciality perennial crops with very low productivity and supply response), PEST-coupled policies do not generally exist. Moreover, for these speciality crops, producer organizations use checkoff schemes to finance joint collective research.

Under alternative configurations of the basic supply and demand parameters, differences in productivity, and the relative political power of producer groups, PERT policies can be over- or undersupplied. Given the joint determination of PERT and PEST policies, however, compensation through various PEST transfers will obviously alter the degree of under- or overprovision relative to the societal optimum.

Analytical Framework

The above historical review demonstrates that the U.S. government is not the perfect benign instrument envisioned by conventional welfare economics. Neither is it, however, the manifestation of powerful interest groups concerned only about their own well-being. In the economic literature, the former extreme presumes that governments engage in the improvement of allocative efficiency through collective action. The latter holds the view that government policy is designed to serve the rent-seekers and politically powerful (Buchanan and Tullock, Bhagwati). In the former instance, all the political power resides in the hands of a benign government that corrects whatever market failures might exist. In the latter, power resides with the interest groups and whatever action is taken by the public sector can be characterized as government failure.

Evidence from the historical record of U.S. agricultural policy rejects both these views. An alternative framework is needed, recognizing that power is distributed between the various interest groups and government—an internally consistent framework that admits the possibility of accommodating these various interests. And it must be acknowledged, moreover, that government does have some separate autonomy. An appropriate model is necessary to conceptualize the bargains and compromises undertaken to shape policy acceptable not only to those who have the greatest capacity to obstruct or stymie the process, but also to others who have stakes in the outcome. Results of the collective process seek to resolve conflict; the consequences can be good, bad, or indifferent for individual sectors of society.

Contrary to much of the literature in economics, the history of U.S. agricultural policy argues for an integrated framework that recognizes the joint determination of both predatory and productive governmental interventions.[5] Many public policies in U.S. agriculture designed to improve allocative efficiency and lower transaction costs in the private sector have also had distributional consequences. To the extent that these consequences are significant, PERT policies will inevitably be linked to the motivational forces that generate purely redistributional PEST policies.

An appropriate integrated framework must not only offer explanatory hypotheses for political economic equilibrium outcomes, but must also be a basis for an operational prescription. This prescription must facilitate the creation of public trust based on a shared sense of legitimate authority. The framework must facilitate the task for public policy economists emphasized by Aaron (p. 13), "... to identify policy rules that are robust and important not only economically but in a fundamental sense politically." In the case of U.S. agricultural policy, what does the identification of policy rules that are politically robust mean? Such robustness can only be examined in this case if we allow for nonseparability of political and economic markets (Rausser, 1982).

One framework for characterizing this nonseparability is provided by the perspective that governments search for the right mix of PERT and PEST policies so as to maximize their political support. Defining s as a total political support, x as the level of PEST interventions, and y as the level of PERT interventions, the problem facing the government is to:

(1) $\quad \underset{x_0, y_0}{\text{Max}} \; s_0(x_0, y_0)$

In the government's attempt to maximize this political support function, interest groups strive to influence the policy selections. Interest groups are formed in order to enhance their social power and to exercise political pressure, and it is hypothesized that there are informed organizations representing each group's interest. They reward the government by extending material benefits and political support when the government's choices further their own interests, and penalize the government by withholding these benefits or by supporting the government's political opponents when these choices are contrary to their interests. This pressure and influence requires that (1) be augmented to include the strengths of various interest groups, i.e.,[6]

(2) $\quad S(x_0, y_0) = s_0(x_0, y_0) + \sum_{i=1}^{n}(c_i, \delta_i),$

where c_i is the "cost of power" to interest group i, and δ_i is an indicator of whether a "reward" or "penalizing" strategy is pursued by interest group i in its attempt to influence government policymakers. Thus,

(3) $\quad p_i(c_i, \delta_i) = \begin{cases} \alpha_i(c_i) & \text{When a reward strategy is selected } (\delta_i = \alpha) \\ -\beta_i(c_i) & \text{When a penalizing strategy is selected } (\delta_i = \beta) \end{cases}$

Note that, while c_i is expressed in terms of the well-being measure for the *i*th interest group, α_i and β_i are expressed in terms of the government's objective function. The first term of equation (2) admits the possibility that the government could be concerned about public interest, the bureaucratic interest, leadership surplus (Froelich, Oppenheimer, and Young, 1971), social welfare, the well-being of a single interest group, or simply the probability of staying elected or appointed as the regulator. Regardless, the government is also composed of individuals with their own private interests. Accordingly, even in a world where the government can be treated as a benevolent pursuer of the public interest, it would be unrealistic to ignore these personal concerns.

For U.S. agricultural policy, decision agents within the government are not oblivious to their own personal material well-being, their social status and political power, and so on. As a result, they respond to potential influence attempts by interest groups who are in a position to reward or penalize, as the second term in equation (2) suggests.

The objective or performance measure criterion function for each interest group is:

(4) $S_i(x_0, y_0) = s_i(x_0, y_0) - c_i$

In some formulations, c_i is treated as political expenditures (Becker, 1983, 1985; Peltzman, 1976), expenditures on organizing and maintaining lobbying efforts (Rausser and de Gorter, 1990), and/or as the cost of political power (Zusman, 1976). Similarly, a number of different specifications may be found in the literature for $s_i(...)$. Peltzman uses an industry profit measure; Becker, in one formulation (1983), uses full income, and later (1985) uses utility measures, which admit more than simply self-interest pursuits by various interest groups; in particular, agents can be altruistic or envious, while empirical studies such as Rausser and Freebairn (1974) and Gardner (1987) use approximate economic surplus measures.

A solution concept must be introduced in drawing out the implication of equations (1) through (4). In Becker's model (1983, 1985), for example, economic interest groups take part in a Cournot-Nash noncooperative game. In the case of Rausser and Foster, each economic interest group behaves as a Stackelberg leader with respect to the government's policy choices; the interest groups set the political terrain over which governments seek to gather the greatest support. Zusman (1976) uses a multilateral bargaining game whose cooperative solution constitutes an equilibrium reflecting various interest groups' social power. Here, we follow the latter formulation, sweeping aside the noncooperative solution, where conflict rules and penalties are imposed by interest groups in their attempts to exert pressure on the government.

The cooperative solution is found by applying the Nash-Harsanyi approach to the simple bargaining game (Harsanyi, 1963). This framework maximizes the product of the differences between the cooperative value of each group's performance measure, (S_i^*) $i=0,1,2,...,n$, and its corresponding initial (formerly defined by Harsanyi as the disagreement value) level $(\tilde{S}_i)$; specifically, the product,

$$(5) \quad U = \prod_{i=0}^{n} [S_i^* - \tilde{S}_i]$$

is maximized with respect to $x_0, y_0, c_1, \ldots c_n$. This, of course, is equivalent to maximizing ln U, since the latter measure is monotone increasing in U. Hence, maximizing

$$(6) \quad \ln U = \ln [s_0(x_0,y_0) + \sum p_i(c_i,\alpha_i) - \tilde{S}_0] + \sum_{i-1}^{n} \ln[s_i(x_0,y_0) - c_i - \tilde{S}_i]$$

results in the following first-order conditions, assuming an interior solution,

$$(7) \quad \frac{\partial \ln U}{\partial x_0} = \frac{1}{S_0^* - \tilde{S}_0} \frac{\partial s_0}{\partial x_0} + \sum_{i-1}^{n} \frac{1}{S_i^* - \tilde{S}_i} \frac{\partial s_i}{\partial x_0} = 0,$$

$$(8) \quad \frac{\partial \ln U}{\partial y_0} = \frac{1}{S_0^* - \tilde{S}_0} \frac{\partial s_0}{\partial y_0} + \sum_{i-1}^{n} \frac{1}{S_i^* - \tilde{S}_i} \frac{\partial s_i}{\partial y_0} = 0,$$

and

$$(9) \quad \frac{\partial \ln U}{\partial c_i} = \frac{1}{S_0^* - \tilde{S}_0} \frac{\partial p_i}{\partial c_i} - \frac{1}{S_i^* - \tilde{S}_i} = 0, \qquad i = 1, ..., n.$$

Multiplying all first-order conditions by $(S_0^* - \tilde{S}_0)$, equations (7) and (8) are also those that would be derived for the maximization of

$$(10) \quad W = s_0(x_0,y_0) + \sum_{i=1}^{n} b_i s_i$$

where

$$(11) \quad b_i = \frac{S_0^* - \tilde{S}_0}{S_i^* - \tilde{S}_i} = \frac{\partial p_i(c_i, a_i)}{\partial c_i}$$

and b_i is presumed to exist and to be a constant.

If the government has all the power, its cooperative and disagreement values will be equal and all b_i will be zero. In this event, if $s_0(\ldots)$ measures the public interest, we have the case of the benevolent government, where only productive behavior will exist ($y_0 > 0$), and with all predatory behavior vanishing ($x_0 = 0$). If interest groups have some power over the government ($b_i \neq 0$), however, then some predatory behavior will exist ($x_0 > 0$), regardless of whether the government internalizes the overall public interest. This is because, as reflected in equations (1) through (4), decision agents constituting a government are not oblivious to their own personal material well-being and their social status, even though the government may be effectively organized to internalize the common group interest.

If one particular interest group has all the social power, the cooperative solution will not exist, nor will the b_i. Essentially, there are two alternative interpretations of the power coefficients, b_i. First, they measure the government's utility gain from cooperation (compared to disagreement), relative to the corresponding gain of the ith interest group. These power coefficients can also be interpreted as the ith interest group's marginal strength of power over the government in a (cooperative) equilibrium (Zusman, 1976).

To complete the structural specification, a representation is also needed for predatory (x_0) and productive (y_0) governmental interventions, and whatever performance measures are used for each s_i. In other words, a representation of the agricultural economic structure

(12) $\quad F(z, x_0, y_0; g)$,

where z is the vector of endogenous variables that form the basis for measuring the s_i ($i = 0, 1, \ldots, n$), and g is a vector of uncontrollable exogenous variables. Combining equation (12) with equations (1) through (10) allows an investigation of the *structural* basis for endogenous agricultural policy.

All empirical studies of endogenous agricultural policy focus only on predatory behavior. Most of the empirical studies specify and econometrically estimate instrument behavioral equations for x_0. These are often defined as policy reaction functions and are the derived decision rules from the structure (1) through (12). Any and all reduced form examinations that focus only on the PEST instruments attempt to isolate, in the choice space, the tangency points of some measure of the possibility frontiers with the indifference curves emanating from the governing criterion (10). Other empirical studies employ revealed preference technology to estimate directly (9); it has been referred to as the governing criterion function (Rausser and de Gorter), the political preference function (Rausser and Freebairn), and the social power objective function (Zusman). Still other empirical studies focus on the economic structure, typically represented by surplus transformation frontiers (Gardner, 1983), and how these frontiers are influenced by the selection PEST instruments. In this setting, the Becker efficient redistribution hypothesis will result in a policy selection that takes the economy to some point on the highest attainable social transformation frontier.[7]

Pressures for Policy Reform

The organizers of this conference asked each of us in the first session to assess the following items: How likely is a major change in current U.S. policy? How important is the GATT Round to the policymakers who have responsibility for the overall economic and trade policies of the United States, compared to those policymakers most concerned about agriculture? Is there agreement or disagreement between the two? The above review of U.S. agricultural policy reveals the many difficulties encountered in attempting to alter existing agricultural policy. Extraordinary circumstances are apparently necessary to make significant reform possible. Indeed, the last successful effort to significantly alter the framework of commodity policy was passed only under the stressful conditions of the Great Depression. Nevertheless, the circumstances characterizing the current state of agriculture have generated speculation that, more than at any time since 1933, significant efforts to reform agricultural policy are likely to achieve some degree of success. This speculation has been fueled by the development of several forces combining to encourage a push for reform in the decade of the 1990s.

The most visible and obvious of these are the recent high budgetary costs of the commodity programs, which rose to unacceptably high levels during the 1980s. These costs dropped back to somewhat more manageable, although still high, levels in 1988 and 1989, largely as a result of the very dry weather experienced in the midwest during the 1988 crop year. The 1988 drought resulted in an abnormally small harvest, a rapid depletion of existing stocks, and consequent high market prices. Combined with the 1985 Farm Bill's increased flexibility in allowing loans to respond to market conditions, the poor crop year of 1988 allowed reduced levels of expenditures in both 1988 and 1989. Given current policies, however, changes in market conditions remained capable of rapidly escalating program costs once again. In a time of great political concern over the national budget deficit, these high levels of program expenditures have generated significant pressure for reform.

Several developments outside of the United States have also contributed to an atmosphere more conducive to policy reform. Agriculture within the United States experienced a boom in the 1970s, due largely to the expansion of agricultural exports. A high growth rate in the world demand for agricultural products made this possible. However, the world demand for agricultural products stagnated in the early 1980s. Further, the dollar strengthened, competition from other exporters became more intense, and U.S. agriculture began to lose its previous market shares (Figure 3.2).

These developments in the export market were directly related to the commodity policies of the major exporting countries. The combination of high levels of output and world prices that were significantly below what domestic producers received for their products, ensured high levels of budgetary expenditures in the United States, the European Economic Community (EEC), and Japan. This combination also required that each employ commercial trade policies to protect their own domestic farm programs. Both the United States and the EEC subsidized agricultural exports and limited the access of imports to their markets, while maintaining coupled transfers to farmers, which kept output high. Additionally, Japan has maintained particularly high levels of effective protection for its producers, who have been supported by the severe limitations placed on imports of many agricultural commodities. These policies led inevitably to increasing tension over trade barriers and subsidies, as each depended upon the expansion of its exports as a vital component in the welfare of its agricultural industries.

These developments in the international arena have taken on great importance because of the Uruguay Round of the GATT negotiations, which began in 1987. The United States and the EEC entered the GATT negotiations facing extremely high farm program budgets. The EEC, concerned with the budgetary problems of its own agricultural programs, adopted production controls for milk for the first time in 1984, and for its other major program commodities in 1987. Consequently, both the United States and the EEC were particularly concerned with reform from the outset of the negotiations, although both entered the GATT rounds burdened by legacies of farm policy that appeared difficult, if not impossible, to change on a unilateral basis.

The difficulty of achieving unilateral policy reform has been due largely to some of the reasons noted above. It has been exacerbated, however, by the fierceness of competition for export markets and the fear that unilateral agricultural policy reform would allow competitors to gain unfair advantage on international markets, while continuing to restrict access to their own markets. This has led to a sort of "prisoner's dilemma" in which many countries feel trapped in their existing policy regimes. Each country fears

Figure 3.2. World Trade in Wheat, Coarse Grains, and Soybeans, 1972-1990 crop years

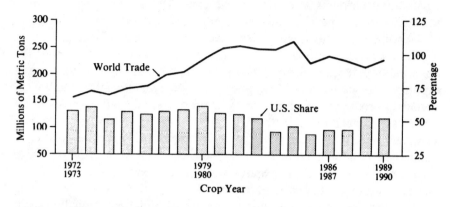

Source: U.S. Department of Agriculture

that unilaterally reducing its own export subsidies, or limiting its own internal level of support to farmers, will result in the loss of market share. The action of one country on its own will rarely induce a significant rise in world prices. Consequently, each country finds that the potential rewards from liberalizing its agricultural policy do not warrant a unilateral move toward reform.

However, simultaneous reform by several important countries might provide an escape from this dilemma. This attitude is characteristic of the participants at the outset of the GATT negotiations, as reflected in the following remarks by Clayton Yeutter (then U.S. trade representative and currently the U.S. secretary of agriculture) in July of 1987: "We are clearly not going to reduce our level of government involvement (in agriculture) unless other people move with us. We are going to go down this road together, and we are going to go down it arm in arm, and we are not going to walk ten steps ahead of the Europeans or the Japanese or anybody else" (Rapp, 1988).

With these factors in mind, the United States and the EEC, as well as the members of the Cairns Group, entered the GATT negotiations with unexpectedly positive attitudes toward reaching a cooperative accord on agricultural trade issues. The accord would include the provision that internal commodity policies designed to protect farmers be lowered multilaterally. Thus, for the first time, participants arrived at the GATT negotiations with a publicly stated willingness to put the components of their own domestic agricultural policies on the negotiating table, a situation that was clear from the outset of negotiations in Punta del Este in 1987. This provision recognizes that domestic governmental interventions are at the core of most distortionary trade policies. The major players' commitment to reach an agreement in the GATT negotiations is a strong reason for optimism about achieving reform of agricultural commodity policy in the early 1900s. If the GATT participants reach an agreement that calls for multilateral reform, they can carry this agreement to each of their respective legislative bodies as a binding commitment. This might provide enough support to overcome the objections of the powerful and entrenched interest groups that will continue to oppose significant reform.

In the United States, Congress already has committed itself to either accepting or rejecting the entire GATT agreement without revision. Therefore, the political battle over accepting the GATT agreement will involve a great number of issues, and battle lines almost certainly will be drawn differently than if the fight dealt only with agricultural issues. For the first time, the coalitions that form to pass or defeat a piece of legislation with a potentially major impact upon agricultural commodity policy will be quite different from those that have ordinarily determined the content of farm bills. This means, of course, that the governing criterion function (10) will be much different for a congressional vote on an entire GATT agreement. This could lead to passage of a GATT agreement that mandates significant reform of domestic agricultural policy. Thus, for several reasons, the GATT negotiations may significantly affect domestic farm legislation.

A final factor justifying optimism about the possibility for reform is a growing belief that the 1990s may well be a decade in which the world demand for agricultural products will again expand at a rapid pace. This development would particularly benefit agriculture if the GATT participants could reach an agreement that would free world markets to accommodate it. Such a turn of events would significantly reduce the adjustment costs that would be experienced by agricultural sectors in the industrial countries as they modify their commodity programs.

Despite the fact that the near budgetary crises of 1986 and 1987 have eased somewhat in the United States and Europe, both continue to issue statements indicating their ongoing commitment to reaching accord in the GATT negotiations. This is promising, as many observers feared that support for the negotiations might erode as budgetary pressures lessened. As their public statements show, however, this has not yet happened.

Current proposals before the GATT call for gradual phaseouts of domestic agricultural programs that couple transfer payments to producers with output levels. Other current proposals seek to eliminate export subsidies and to convert all import restrictions to bound tariffs. These measures, by improving access to import markets and by fostering freer competition in export markets, are intended to orient domestic production more effectively to market forces. Current proposals also call for improved and standardized sanitary and phytosanitary measures specifically aimed at protecting animal, plant, and human health. If adopted, these measures would bring agriculture into line with GATT regulations imposed upon other industries. They would also allow the governments involved to take the agreement to their respective legislative bodies in the fast-track manner suggested above.

Congress will probably pass the 1990 Farm Bill before the current GATT negotiations conclude. If this happens, existing farm policy will undergo few substantive changes, and a slightly modified version of the 1985 bill will pass without much struggle. The agricultural policymakers will then focus their attention on the outcome of the GATT round. They will have to present the mandates of the final GATT agreement before Congress, who then must draft special legislation to generate a domestic agreement that brings internal policy into compliance with the GATT agreement.

The pieces of the puzzle are ready to fall into place. Policymakers appear to be in a historically unique position that offers the potential for momentous courses of action. Agricultural policy is at a critical point—some would argue, a potential turning point—as we head into the 1990s. Significant action, however, may not be taken. Whether policymakers at GATT and in the legislative bodies of the participating countries will seize the moment of opportunity before them is far from certain.

In any event, developments in commodity policy will be largely determined in Geneva, through GATT, and not in Washington. The protestations of the commodity organizations notwithstanding, significant reform directed at increased market orientation for agriculture will likely benefit U.S. agricultural producers. U.S. agriculture will probably fare well in a more competitive environment in which technological innovation, value added, and product quality become increasingly important. Further, optimistic forecasts regarding the global economic outlook signal the possibility of expanded world demand for agricultural products, a demand which will likely be enhanced and more readily accessed in a freer trade environment. Thus, significant reform of U.S. agricultural commodity policies is likely to benefit not only consumers and taxpayers, but also U.S. agriculture as a sector.

While the potential for change in commodity policy attracts much attention, a number of other increasingly important issues will also confront agricultural policymakers in the 1990s. Policymakers will almost certainly address these issues regardless of the fate of commodity programs. In some cases, new issues have arisen as a result of technological progress and the continuing development of domestic and global economies. In other cases, they have arrived on the agenda as a result of new understandings of the way in which policy does, and might do, its work.

An area certain to require increased attention from agricultural policymakers in the 1990s is the rapid development of new biotechnologies, which are advancing particularly rapidly in the animal sciences. The bovine growth hormone, which dramatically increases the milk production of dairy cows, is only one of many important examples. The plant sciences also continue to make headway in this area. These developments will push out the supply curves for agricultural commodities and will reinforce the long-standing trend toward fewer and larger farms. The continued development of these technologies will necessitate and will depend upon increased government involvement in providing appropriate regulation, establishing guidelines to determine property rights, monitoring the transfer of technologies, and providing public support for necessary research activities.

The development of the new technologies in agriculture as well as in other sectors of the economy is closely related to the increasing political importance of environmental and health issues. Advocacy groups representing these concerns should enjoy increasing influence in the policy forum. These groups have become better organized and have benefited from "learning by doing" during the 1985 Farm Bill debate. Several specific issues that are expected to arise in this context include the problems of groundwater contamination, coastal water pollution, soil erosion, and persistent chemical residues in the soil and agricultural products. As previously noted, each of these specific environmental problems is linked indirectly to the nature of the coupled commodity programs. It will become increasingly important to understand these types of linkages between the effects of otherwise separate programs in order to accurately evaluate the true costs and benefits of specific programs and to construct optimal policy mixes.

Another issue likely to receive increasing attention is the enormous impact of microeconomic variables and events upon the agricultural industry. The 1990s will likely see an increasing emphasis on protecting agriculture from the damaging effects of short-term fluctuations in the macroeconomy. This issue will be dealt with in part by reforming government-sponsored enterprises in agriculturally oriented credit markets. The linkages between agriculture and the macroeconomy are closely related to the importance of the international marketplace and the consequent impact on the agricultural sector of exchange rates, international capital markets, and the trade environment. These linkages currently receive a great amount of attention, and almost certainly will find increased representation in formulating future agricultural programs.

Finally, more careful and useful ways should be found to distribute income assistance in the agricultural sector. Income-assistance in past programs to primarily large and relatively high-income recipients will be refined in future farm legislation.

Concluding Remarks

The integrated productive and predatory framework outlined in this chapter not only offers an explanation for differing political economic equilibriums that have arisen over the years in U.S. agriculture, but also provides the basis for reform. In coordinating both PEST and PERT policies through design and implementation, conventional economic frameworks must be swept aside. Instead, formulations must be developed that explicitly recognize the level and distribution of political power and influence. In this setting, prescriptive analysis can facilitate the search for a consensus on the trade-off between public and personal interests that is both economically and politically robust.

Operational prescription must recognize not only the economics of various policies, but also the level and distribution of political power as reflected in equation (10). There are a number of ways to motivate changes in the distribution and level of political power so that policies will be more productive and less predatory. These include being in a position to design and implement creative and productive packages when the timing is ripe [a major external crisis reflected by significant changes in (12)]; developing an innovative political technology that allows adoption of a new mix of x_0 and y_0; and changing the institutional structure or the constitution, effectively altering the cost of power and its exercise through threat and reward strategies [equations (2) and (3)].

As argued in the section "Pressures for Policy Reform," above, an excellent example of institutional changes that may alter the level and distribution of power arises from the ongoing GATT negotiations. In the current Uruguay Round, all parties recognized and accepted that distortionary trade policies exist in agriculture to rationalize internal policies, and thus that both sets of policies should be included in the negotiations. Regardless of whatever negotiated settlement emerges from Geneva, Congress has already committed itself to either accepting or rejecting it without revision. This means, of course, that the interest-group configuration reflected in equation (10) after the completion of the GATT negotiations could be dramatically different from the political landscape that has existed over much of the prior fifty or so years. To be sure, the vote on any particular GATT agreement could vary widely from the political economic structures that have historically determined the content of past U.S. agricultural policy. The coalition of interest groups likely to determine whether a GATT agricultural code is accepted or rejected will not be dominated by the same interests that have prevailed in past debates over farm legislation. If the GATT agreement is accepted by the U.S. Congress, the executive branch will no doubt agree, and consequently impose an external code that will mandate some reform of current domestic agricultural policy.

Having outlined a number of ways in which agricultural policy might evolve and change in the 1990s, it is appropriate to acknowledge that government regulation will continue to be pervasive in agriculture, regardless of how particular rounds of legislation unfold. The major uncertainties lie in the degree of protection the government will continue to afford producers through commodity market intervention, and in the degree of legislative response generated by emerging issues, such as concern for the environment and health-related matters.

Notes

1. This was implemented through a nonresource loan program where the loan rate became, in effect, the price support.

2. Note that, in a world of limited information, price support transfers can dominate pure lump sum transfers (Foster and Rausser, 1989). Innovative producers, who would otherwise lose as a result of adopting the technology emerging from public investment in research, are offered compensation through coupled price supports. As a result, these innovative producers defect from interest group coalitions that attempt to obstruct the public investment in research or the adoption of any technology that emerges from such investments. All producers receive some compensation, but those who innovate receive a higher level. This result is consistent with the observation of Blackorby and Donaldson (1988) that the "superiority of transfers of purchasing power over transfers of goods and services disappears" when information is limited. The reason for this result is that transfers based on personal characteristics will induce persons without those characteristics

to either mimic the intended recipients and fool the transferring agency, or to actually adopt those characteristic s.

3. The secretary of agriculture was given discretion in the case of feedgrains and wheat to lower the price support up to 20 percent below the basic loan rate. For soybeans, the loan rate can be lowered no more than 5 percent. For cotton and rice, the effective loan rate is set at world market prices. As a result, cotton and rice farmers participating in government programs can first pledge their output as collateral for a loan at the basic rate and, at maturity, repay the loan at the prevalent world market price if it is lower than the basic rate.

4. The land base for each producer is established at the county office of the Agricultural Stabilization and Conservation Service (ASCS). The ASCS is the administrative agency within the U.S. Department of Agriculture that has responsibility for implementation. It has an office in each state and 3,000 county offices nationwide. In addition to several thousand employees, a local committee of three persons (usually producers) handles local appeals of decisions and other administrative matters. County offices assign each local producer a program yield as well as a land base.

5. The best the economic literature has been able to offer in the design of democratic decision making frameworks is to *separate* the processes for each of the two types of policies: PERTs and PESTs. Long ago, Wicksell (1896) recognized the distinction between the two and argued for organizing government so that the provision of the two policies would be decided by separate and qualitatively different processes. Mueller (1989), in his recent survey of the literature, outlines the conceptual and practical advantages of considering the two separately, noting that joint determination results in outcomes not generally characterized by unique or stable equilibria.

6. To simplify the analysis, an additive strength function is imposed. For the case of a more general specification, see de Gorter, Nielson, and Rausser; and Rausser and de Gorter.

7. The joint determination of x_0, y_0 is analytically examined in Rausser and Foster, and in Rausser (1990). A number of extensions of the basic formulation (1)–(12) may be found in Rausser (1990). These extensions include: uncertainty and limited information, dynamics incorporating inconsistent planning horizons between special and public interests, the policy proliferation process, and the organizational costs of commodity interest groups.

References

Alston, J.N., W.E. Edwards, and J.W. Freebairn. "Market Distortions and Benefits From Research." *American Journal of Agricultural Economics,* 70 (1988), pp. 281-88.

Becker, G.S. "A Theory of Competition Among Pressure Groups for Political Influence." *Quarterly Journal of Economics*, 58 (1983), pp. 371-400.

———. "Public Policies, Pressure Groups, and Dead Weight Costs." *Journal of Public Economics*, 28 (1985) pp. 329-47.

Bhagwati, J. "Directly Unproductive, Profit-Seeking (DUP) Activities." *Journal of Political Economy*, 90 (October 1982), pp. 998-1002.

Blackorby, C., and D. Donaldson. "Cash Versus Kind, Self-Selection, and Efficient Transfers." *American Economic Review*, 78 (September 1988), pp. 691-700.

Bosso, C.J. *Pesticides and Politics: The Life Cycle of a Public Issue*. Pittsburgh: University of Pittsburgh Press, 1987.

Brandow, G.E. "Part III, Policy for Commercial Agriculture 1945-71." In Martin, L.R., ed. *A Survey of Agricultural Economics Literature*, Vol. 1. Minneapolis: University of Minnesota Press, 1977.

Buchanan, J.M., and G. Tullock. *The Calculus of Consent*. Ann Arbor: University of Michigan Press, 1962.

Carson, R. *Silent Spring*. New York: Fawcett Publications, Inc., 1962.

de Gorter, H., D. Nielson, and G. Rausser. "The Political Economy of PERTs and PESTs in U.S. Agriculture." Department of Agricultural and Resource Economics, University of California at Berkeley. Unpublished paper, 1990.

Englebert, E.A., and A.F. Scheuring, eds. *Competition for California Water: Alternative Resolutions*. Berkeley: University of California Press, 1982.

Ericksen, M.H. "The Use of Land Reserves to Control Agricultural Production." Publication ERS-635. Washington, D. C.: USDA, ERS, September 1986.

Foster, W.E. "Agricultural Policy and Commodity Supply Analysis." Ph.D. Dissertation, Department of Agricultural and Resource Economics, University of California at Berkeley, 1987.

Foster, W.E., and G.C. Rausser. "A Political-Economic Rationale for Coupled Welfare Transfer Policies." Working Paper No. 498, Department of Agricultural and Resource Economics, University of California at Berkeley, 1989.

———. "Farmer Behavior Under Risk of Failure." *American Journal of Agricultural Economics*, forthcoming, 1990.

Froelich, N.J., A. Oppenheimer, and J. Young. *Political Leadership and Collective Goods*. Princeton, New Jersey: Princeton University Press, 1971.

Gardner, B.L. "Efficient Redistribution Through Commodity Markets." *American Journal of Agricultural Economics*, 65 (1983), pp. 225-34.

———. "Causes of U.S. Farm Commodity Programs." *Journal of Political Economy*, 95 (1987), pp. 290-310.

Harsanyi, J.C. "A Simplified Bargaining Model for the End Person Cooperative Gain." *International Economic Review*, 4 (1963), pp. 194-220.

Hochman, E., and D. Zilberman. "Examination of Environmental Policies Using Production and Pollution Microparameter Distributions." *Econometrica*, 4 (1978), pp. 739-60.

Holmes, B. H. "History of Federal Water Resources Programs and Policies, 1961-70." Misc. Publication No. 1379, Washington, D.C.: USDA, 1979.

Krueger, A. D. "The Political Economy of the Rent-Seeking Society." *American Economic Review*, 64 (1974), pp. 291-303.

Lichtenberg. E., and D. Zilberman. "The Welfare Economics of Price Supports in U.S. Agriculture." *American Economic Review*, 76 (1986), pp. 1135-41.

Love, A., and W. Foster. "Commodity Program Slippage Rates for Corn and Wheat." *Western Journal of Agricultural Economics*, forthcoming.

Lowi, T.J. "How Farmers Get What They Want." In Lowi, T.J., ed. *Legislative Politics USA*. New York: Little, Brown, Inc., 1965.

Macintyre, A.A. "Why Pesticides Received Extensive Use in America: A Political Economy of Agricultural Pest Management to 1970." *Natural Resources Journal*, 27 (Summer 1987), pp. 533-78.

Mitchell, R.C. "National Environmental Lobbies and the Apparent Illogic of Collective Action." In Russell, C.S., ed. *Collective Decision Making: Applications From Public Choice Theory.* Baltimore: Johns Hopkins University Press, 1979.

Mueller, D.C. *Public Choice II.* New York: Cambridge University Press, 1989.

Norton, N.W. "The Effect of Acreage Reduction Programs on the Production of Corn, Wheat, and Cotton: A Profit Function Approach." Presented at the American Economics Association annual meetings in Reno, Nevada, July 1986.

Pavelis, G.A. "Natural Resource Capital Formation in American Agriculture: Irrigation, Drainage, and Conservation, 1955-1980." Washington, D.C.: USDA, ERS, 1985.

Peltzman, S. "Toward a General Theory of Regulation." *Journal of Law and Economics*, 211 (1976), 211-40.

Perkins, J. *Insects, Experts, and the Insecticide Crisis.* New York: Plenum Press, 1982.

Rapp, D. *How the U.S. Got into Agriculture and Why It Can't Get Out.* Washington: Congressional Quarterly, 1988.

Rausser, G.C. "Political Economic Markets: PESTs and PERTs in Food and Agriculture." *American Journal of Agricultural Economics*, 64 (December 1982), pp. 821-33.

―――― "Predatory Versus Productive Governments: The Case of U.S. Agricultural Policy." *Journal of Economic Perspectives,* forthcoming, 1990.

Rausser, G.C., and H. de Gorter. "Endogenizing Policy in Models of Agricultural Markets." In Maunder, Alan, and Alberto Valdez, eds. *Agriculture and Governments in an Interdependent World.* Oxford: Oxford University Press, forthcoming.

Rausser, G.C., and J.W. Freebairn. "Estimation of Policy Preference Functions: An Application to U.S. Beef Import Policy." *Review of Economics and Statistics*, 56 (1974), pp. 437-49.

Rausser, G.C., and R. Howitt. "Stochastic Control of Environmental Externalities." *Annals of Econ and Social Measurement*, 4 (1975), pp. 227-92.

Rausser, G. C., and H. Lapan. "Natural Resources, Goods, Bads, and Alternative Institutional Frameworks." *Resources and Energy*, 2 (1979), pp. 293-324.

Reichelderfer, K. "Do USDA Farm Programs Contribute to Soil Erosion?" Ag. Econ Report No. 532, Washington, D. C.: USDA, 1985.

Roningen, V.O., and P.M. Dixit. *Economic Implications of Agricultural Policy Reform in Industrial Market Economics.* USDA Staff Report, AGES 89-36, Washington, D.C.: USDA, 1989.

Ruttan, V.W. *Agricultural Research Policy.* Minneapolis: University of Minnesota Press, 1982.

Tweeten, L. "Adjustments in Agriculture and Its Infrastructure in the 1990s." In *Positioning Agriculture for the 1990s: A Decade of Change.* Washington, D.C.: National Planning Commission, 1989.

Tyers, R., and K. Anderson. "Distortions in World Food Markets: A Quantitative Assessment." Washington, D.C.: Paper prepared for the World Bank, *World Development Report 1986.*

Wicksell, K. "A New Principle of Just Taxation." *Finanztheoretische Untersuchungen*, Jena, 1896. Reprinted in Musgrave and Peacock, eds. *Classics in the Theory of Public Finance.* New York: St. Martin's Press, 1967.

Worster, D. *Rivers of Empire: Water, Aridity and the Growth of the American West.* New York: Pantheon Press, 1985.

Zietz, J., and A. Valdez. "The Costs of Protectionism to Developing Countries." World Bank Staff Working Papers No. 769, Washington, D.C.: The World Bank, 1986.

Zusman, P. "The Incorporation and Measurement of Social Power in Economic Models." *International Economic Review*, 17 (1976), pp. 447-62.

Zusman, P., and G. Rausser. "Organizational Equilibrium and the Optimality of Collective Action." Working Paper No. 528, Department of Agricultural and Resource Economics, University of California at Berkeley, 1990.

SECTION TWO

The International Agricultural and Trading Environment

4

Structural Change in Canadian, United States, and European Agriculture

George L. Brinkman

Introduction

Global agriculture underwent substantial changes during the 1980s. The growth-oriented conditions of the 1970s altered significantly in the 1980s, with growing international competition, trade wars, declining international prices, and increasing pressures for government intervention and assistance. In North America, both Canadian and U.S. farmers have experienced declining incomes and falling land values, generating concerns about the future of the family farm and the structural attributes associated with this form of business operation. In the European Community (EC), protectionist policies and high support prices have insulated farmers from fluctuating world price levels while encouraging increased production. Concern for farming and farm structure has also arisen in Europe, however, not only from the standpoint of maintaining a viable farming population, but also from the focus of formulating social policy in support of rural development and environmental preservation.

This chapter provides an assessment of recent structural changes in Canada, the U.S., and the European Community. Farm structure relates to a variety of factors that affect the composition, organization, and operation of farms. At the farm level, these include the number, size, and distribution of farms, capital investments, form of business operation, tenure, off-farm work, hired labor, and financial characteristics. At the macro/sectoral level, these also include productivity, product and input price ratios, and self-sufficiency levels.

The chapter is organized into four sections. The first summarizes the major components (or lack thereof) of structural policies in Canada, the

U.S., and the European Community. The second deals with the major structural dimensions and changes of structural conditions of individual farms, while the third summarizes macro/structural characteristics of the agricultural sectors in the various countries. The final section provides the conclusions and implications for policy.

Farm Structure Policy

The structure of agriculture in any country is determined by a number of items: physical, climatic, and human resources; historical rights and developmental patterns; attitudes toward property ownership and use; and many other factors, including external influences and conditions of trade. Some countries may choose to develop specific structural policies to guide the size, distribution, composition, and organization of farms, while others may take a less interventionist approach and let farm structure evolve in response to other marketing and production policies and opportunities.

In Canada, the structural organization of agriculture has not been as prominent a public policy issue as in Europe, nor has structural policy been a central pillar of Canadian farm policy. To be sure, concerns with structural aspects of farming have been present in Canada, as is evidenced by both public attitudes and the content of farm programs. One need only cite the uniformly favorable view taken by policy makers, farm organizations and the public at large of the family farm as the basic economic unit of agriculture and the central unit of rural society. There are also innumerable examples of structural concerns evident in agricultural programs, such as the special treatment of agriculture under the tax code; the formation of producers' marketing boards to halt the integration of farm production by corporate agro-industries; the farm enlargement and consolidation features of the former Small Farm Development and ARDA programs; special farm assistance programs designed to save the family farm; continuing efforts to define farms and farmers to determine program eligibility; limits on farm assistance based on farm size; and various programs aimed at helping new farmers gain access to the industry. Furthermore, agriculture has been used as a vehicle for regional development, serving in a limited capacity as the focus of funds designed to improve economic conditions in lagging regional areas. Thus, it is not suggested that structural matters have been absent from Canadian farm policy, but structural change in agriculture has not been a major focal point of public debate about the industry. More importantly, structural policy has not been an explicit and coherent component of Canadian agricultural policy development.

In the United States as well, structural issues have traditionally been addressed through commercial agricultural and development policies, rather than through specific structural policies. As in Canada, U.S. farm structure has focused to a large extent on the maintenance of the family farm. It has also been influenced by a variety of taxation, credit, and commercial agricultural policies, as well as land reclamation and development policies (particularly in the West). While many of these policies have had maximum eligibility criteria, many have also been subject to widespread manipulation (through multiple family member ownership, etc.) whereby farms have increased substantially in size, while retaining their eligibility for program benefits. As a result, farm structure in the U.S. has emerged with a strong emphasis on the achievement of efficiency and competitiveness, with more involvement of larger-than-family ownership and control by agribusiness than found in Canada.

The importance of farm structure in the agricultural policy of the United States has fluctuated widely over time. During the 1970s, there developed in the United States a sense that the structural transformation of American agriculture had perhaps gone too far; the economic organization of agriculture was departing too much from the family farm structure that was preferred by the public at large, by farmers themselves, and by policy makers. This perception animated a major study of structural change in U.S. agriculture undertaken by the USDA under the leadership of Secretary of Agriculture Bergland in the late 1970s. The final report of that study, *A Time to Choose* (1981), together with other reports prepared in the same period (Shertz et al., *Another Revolution in U.S. Agriculture*, 1979; USDA, *Structure Issues of American Agriculture*, 1979), heightened awareness of farm structural issues throughout the country. The findings of these reports, however, never became prominently translated into structural policy; a change in the federal administration brought about a de-emphasis of structure in agricultural policy and a renewed emphasis on efficiency. Since the early 1980s, interest has increased in agriculture's role in rural development, but the major concern has been with farm financial difficulties, international competitiveness, and trade policy reform.

In Europe, farm structure has traditionally received more attention, as farm policy has had a long history of supporting small-sized operators and actively directing farm structure. In the early 1960s, structural policy in a number of continental European countries focused on land consolidation, as years of land subdivision among multiple farm sons left many farm holdings fragmented and spread out in hard-to-service parcels. With the development of the European Economic Community, a specific Agricultural Structures Policy was developed to promote the modernization of farms through financial assistance, to create incomes for farmers approaching that earned by nonfarmers (in manufacturing), to provide inducements for early retirement to free up land for farm consolidation, and (to a lesser extent) to provide socio-economic guidance and vocational training. By 1975, provisions were also implemented to provide direct income support to farmers in disadvantaged (mountain and less favored) areas—in essence, to compensate for natural handicaps and maintain farm holdings in these areas. Further provisions were added in 1977 and 1979 to assist in agricultural processing and regional development. Mounting surpluses in the early and mid-1980s, however, caused the Economic Community to reassess its policies, resulting in a new emphasis in the late 1980s on achieving a better market balance, as well as maintaining viable rural communities and promoting environmental protection and conservation.

While structural dimensions have not been an explicit focal point of agricultural policy in either Canada or the U.S., there are several reasons to believe that they will continue to be important considerations in the future. Among the reasons are the following:

- It is recognized that the strengthening of the agricultural sectors in the two countries may require further changes in the economic organization of agriculture, and restructuring of both the technical and business aspects of farming.
- In times of budgetary restraints, the best use of limited public funds may require that programs be tailored to the particular situations and needs of specific target groups within agriculture.
- As farming becomes more industrialized and more closely integrated with the general economy, and as resource use and output in agriculture become more concentrated, general economic conditions (e.g., growth rates, inflationary trends, the operation of capital markets, interest rates, investment incentives, taxation policies, and exchange rates) impact more directly on the commercial sector of the industry. Farmers and agricultural policymakers must take more account of the effect of these variables on the industry's structure and performance.
- The structure of agriculture is changing. In particular, structural diversity is increasing, with significant departures from the traditional stereotype of an industry composed of relatively homogeneous "family farms." This has two consequences. First, it becomes progressively more difficult to rationalize an undifferentiated policy

posture toward agriculture based on conditions that no longer exist. Policy must increasingly acknowledge the structural heterogeneity of the industry. Second, there is no reason to believe that society is indifferent about the structure of farming. Accordingly, public policies directed at agriculture, or which impinge upon it, might contain structural engineering elements or, at a minimum, might seek to avoid the inadvertent creation of an agricultural structure and performance that is incompatible with societal preferences and that no one, inside or outside of agriculture, really wants. In particular, we are likely to see increased societal concern with such "green" issues as food safety, environmental preservation, and animal welfare.

Finally, within the framework of the current GATT negotiations, domestic agricultural policies have been incorporated into the negotiations, along with more traditional "trade" policy measures. This means that potential trade agreements and subsequent adjustments in domestic agricultural policies could be both influenced by, and directly influence, structural conditions in the various negotiating countries.

Structural Conditions in Canada, the U.S., and the European Economic Community

Agriculture in Canada has developed under a number of conditions similar to those found in the United States,[1] so it is not surprising that both countries have quite similar structures. There are some differences, however, in both farm characteristics and policy. Farm structure in Europe, on the other hand, has evolved through a much longer history of land cultivation, with a greater preponderance of small-holder operations. As a consequence, differences between Europe and North America are much greater than between Canada and the United States, especially with respect to farm size.

Size and Distribution of Farms by Land Area

Table 4.1 provides a summary of farm numbers and average size in hectares for Canada, the U.S., and Euro 10 Economic Community for selected years from 1960/61 to 1985/86. For Canada, farm numbers have declined from 481 thousand farms in 1961 to 293 thousand in 1986, with a corresponding increase in average farm size from 145.3 hectares to 231.5

hectares. The corresponding figures for the U.S. from 1960-1985 show a decline from 3,955 thousand to 2,293 thousand farms, with an increase in average size from 119.8 to 179.3 hectares. The Euro 10 figures show a decline from 8,147 thousand to 5,104 thousand farms, and an increase in average size from 11.2 to 17.3 hectares. It is quite apparent that farms in the European Community have remained, on average, much smaller than those in Canada and the U.S., but the *rate of change* in both the decline in farm numbers and increasing average farm size have been fairly similar in the three areas.

Tables 4.2 to 4.4 summarize the distribution of farms by size in acres/hectares for Canada, the United States, and the Euro 10 Economic Community. The Canadian data show a gradual increase in the number of farms with 760 acres or more, while the U.S. data show steady or

Table 4.1 Number of Farms and Average Farm Size, Canada, U.S., and the Euro 10 Economic Community, Selected Years, 1960-86

Year	Canada		U.S.		European Community	
	1000 Farms	Average Hectares per Farm	1000 Farms	Average Hectares per Farm	1000 Farms	Average Hectares per Holding
1960			3955	119.8	8147	11.2
1961	481	145.3				
1970			2944	150.9	6588	14.0
1971	366	187.4				
1975			2517	170.0	5900	15.3
1976	339	201.9				
1980			2428	172.8	5496	16.2
1981	318	206.8				
1985			2293	179.3	5104	17.3
1986	293	231.5				
1985/86 percentage of 1960/61	61	159	58	150	63	154

Sources: Canadian data from Statistics Canada, *Census of Agriculture*, various years; U.S. data from USDA, *Farm Real Estate: Historical Series Data, 1950-85*; and European Community data from Eurostat, *Agriculture Statistical Yearbook*, 1989.

Table 4.2 Farm Numbers and Percentage Distribution by Size in Acres, Canada, Selected Years, 1961-86

	1961		1971		1981		1986	
Size Group in Acres	1000 Farms	%	1000 Farms	%	1000 Farms	%	1000 Farms	%
Farms classified by size of farm								
Under 10								
10 – 69	50	3.5	14	3.9	16	5.2	14	5.0
70 – 239	204	42.5	128	34.8	99	31.1	87	29.7
220 – 399	83	17.2	59	16.4	47	14.8	43	14.6
400 – 599	45	9.3	36	9.8	28	8.7	25	8.6
560 – 759	32	6.6	29	7.9	24	7.5	22	7.5
760 – 1,119	28	5.7	30	8.2	28	8.7	26	9.0
1,120 – 1,599	13	2.7	17	4.6	18	5.7	19	6.3
1,600 and over	10	2.0	14	3.9	18	5.6	21	7.2
Total	481	100.0	366	100.0	318	100.0	293	100.0

Source: Statistics Canada, *Census of Agriculture*, various years.

Table 4.3 Farm Numbers by Size in Acres, United States, Selected Years, 1959-87

Size Group in Acres	1959	1969	1982	1987
		Thousand Farms		
Small:				
Under 10	244	162	187	183
10–49	813	473	449	412
Medium:				
50–99	658	460	344	311
100–179	773	542	368	334
180–259	415	307	211	192
Large:				
260–499	472	419	315	286
500-999	200	216	204	200
1000 and over	136	151	161	169
Total	3711	2730	2239	2088
Average size/farm	303	389	440	462

Source: U.S. Dept. of Commerce, *Census of Agriculture*, various years.

Table 4.4 Farm Numbers by Size in Hectares of Agricultural Area Used (AA), Euro 10 Economic Community, Selected Years, 1960-85

Size Group in Hectares	1960	1970	1975	1980	1985
	\multicolumn{5}{c}{Thousands of Farm Holdings}				
1 – <10	5742	4332	3772	3462	3140
10 – <20	1329	1116	936	839	761
20 – 50	820	850	867	852	821
50 or more	265	291	325	343	382
Total	8147	6588	5900	5496	5104

Source: Eurostat, *Agriculture Statistical Yearbook*, 1989.

increasing numbers with 500 or more acres. European data show relatively constant numbers (but an increasing percentage) of farms in the 20-50 hectare size range and an increasing number in the 50+ hectare category.

Size and Distribution of Farms by Economic Class

Tables 4.5 to 4.7 examine farm size in terms of gross sales for Canada and the U.S., and in terms of gross margins for the Euro 10 Economic Community. These tables compare changes in the distribution of both the number of farms and the percentage of total gross farm sales/margins by different gross sales/margin categories. By showing both changes in farm numbers and sales/margins by different size categories, the tables also provide some measures of concentration.

In Canada, most of the farms have been distributed in the smaller economic size categories, but most of the sales are earned by the farms in the larger size ranges. Furthermore, both the share of farms and the share of sales in the larger categories have been increasing over time. In 1976, only 3.6 percent of all Canadian farms had gross sales of $100,000 or more, accounting for 30.8 percent of all sales. By 1986, the distribution of farms in the same category had grown to 19.9 percent and the share of gross sales to 67.6 percent. In contrast, the percentage of farms with gross sales under $10,000 declined from 45.3 to 28.5 percent over the same period, while their share of gross sales fell from 6.2 to 1.6 percent.

The corresponding figures for the U.S. cover an eighteen-year period as opposed to a ten-year period for Canada, but they show a relatively similar pattern. Farms with gross sales of $100,000 or more in 1970 accounted for 1.7 percent of all farms and 31.5 percent of all gross sales, while those with under $10,000 in gross sales accounted for 70.1 percent of all farms and 17.0 percent of sales. By 1988, the largest farms accounted for 15.7 percent of all farms and 70.9 percent of all sales, while those with under $10,000 in gross sales accounted for 46.9 percent of all farms but only 5.5 percent of all sales.

Compared to Canada, the distribution of farms and gross sales in the U.S. shows a lower percentage of farms with gross sales over $100,000, but a higher percentage of output by farms in the larger gross sales categories (19.9 percent of all farms producing 67.6 percent of all output in Canada, compared to 15.7 percent of all farms producing 70.9 percent of all sales in the U.S.). At the small size of the spectrum, the U.S. data also show a higher proportion of smaller farms than in Canada. Overall, these data indicate that Canada has a more even distribution of both farms and gross sales across the various size classes, with U.S. data showing a lower proportion (and greater decrease since the 1982 census) of both farms and sales in the middle sized categories. In Canada, increases in farm size have been relatively similar across different size classes for the last fifteen years, with the largest 1, 5, and 25 percent of farms producing 18, 37, and 74 percent respectively of all gross sales in 1986, compared to a similar 18, 37, and 72 percent respectively in 1971.

Table 4.5 Percentage Distribution of Farms and Gross Sales by Economic Class, Canada, 1976, 1981, and 1986

Year	Sales Class ($ Canadian)						
	Under 5,000	5,000 9,999	10,000 24,999	25,000 49,999	50,000 99,999	100,000 249,999	250,000 or more
	Distribution of Farms						
1976	31.8	13.5	24.1	17.5	9.2	—3.6—	
1981	23.4	10.7	18.5	18.6	17.1	9.1	2.3
1986	18.6	9.9	16.9	16.0	18.7	15.4	4.5
	Distribution of Gross Sales						
1976	2.4	3.8	15.3	23.7	24.0	—30.8—	
1981	0.9	1.5	6.2	13.4	23.9	26.6	27.4
1986	0.6	1.0	3.9	8.1	18.9	32.3	35.3

Source: Statistics Canada, *Census of Agriculture*, 1976, 1981, 1986.

Table 4.6 Percentage Distribution of Farms and Gross Sales by Economic Class, U.S., 1970, 1980, and 1988

Year	Under 5,000	5,000 9,999	10,000 24,999	25,000 49,999	50,000 99,999	100,000 249,999	250,000 or more
				Sales Class ($ U.S.)			
				Distribution of Farms			
1970	57.5	12.6	12.3	10.2	5.6	1.2	0.5
1980	38.2	12.8	11.8	11.6	14.5	6.8	4.3
1988	34.2	12.7	12.5	11.4	14.6	9.8	5.9
				Distribution of Gross Sales			
1970	10.2	6.8	12.1	18.7	20.7	10.1	21.4
1980	4.2	2.7	4.0	6.9	18.5	18.5	45.0
1988	3.2	2.3	3.5	5.5	14.6	21.9	49.0

Source: USDA, *Economic Indicators of the Farm Sector, National Financial Summary, 1988*.

Within the European Community, data in Table 4.7 for the Euro 10 show a pattern of distribution and change similar to that within Canada and the U.S., although the data are not quite comparable. The European data show an increasing concentration of farms and gross margin in the larger gross

Table 4.7 Percentage Distribution of Farms and Gross Margin by Economic Gross Margin Class, Euro 10 Economic Community, 1975, 1980, and 1985

Year	< 2	2–3.9	4–5.9	6–7.9	8–11.9	12–15.9	16–39.9	40–99.9	100+
				Level of Standard Gross Margin in ESU*					
				Distribution of Farms					
1975	44.4	17.0	9.3	6.4	8.3	5.0	8.0	1.4	0.2
1980	44.3	14.0	7.7	5.1	6.8	4.7	12.4	3.8	0.8
1985	33.9	15.6	8.7	6.0	7.9	5.1	14.6	6.7	1.5
				Distribution of Standard Gross Margin					
1975	6.1	7.6	7.3	7.1	13.1	10.9	28.7	12.1	7.2
1980	3.0	4.3	4.1	3.8	7.0	6.7	30.9	24.5	15.8
1985	2.2	3.3	3.2	3.0	5.5	5.1	27.1	28.6	22.0

*In 1980, 1 ESU = 1000 ECU = $1310 U.S. or $1565 Canadian.
In 1985, 1 ESU = 1100 ECU = $ 977 U.S. or $1365 Canadian.
Source: Eurostat, *Agricultural Statistical Yearbook*, 1989.

margin size classes over time, but the farms tend to be smaller than those in North America (i.e., a gross margin of 100 ESU in 1985 was the equivalent of about $136,000 Canadian, or somewhere around $300,000-$400,000 in gross sales).

Changes in Farm Size and Distribution by Constant Dollar Sales Classes

Although the preceding tables show substantial increases in the number of farms in the larger size categories over time, it must be recognized that some of this growth is due to inflation rather than increased physical output. As a consequence, when making comparisons of size and distribution over time, it is important to correct for the effect of product price increases, in order to avoid the appearance of growth simply from rising prices, rather than increased volumes of real production.

Tables 4.8 and 4.9 provide a comparison of Canadian farm size distributions using nominal and real gross sales categories of size. Table 4.8 shows the number of farms with gross sales of $50,000 or more measured in nominal terms growing from 2.9 percent of all farms in 1971 to 38.6 percent in 1986, a 13.3-fold increase in the percentage level. Unfortunately, a significant share of this increase is due to farms growing larger not because of real growth but because of inflation in product prices. In real terms (Table 4.9), farms also showed an increase in size, but the percentage of farms with gross sales of $50,000 or more increased only from 9.2 to 29.0 percent—a 3.2-fold increase. By 1986, farms with 1975

Table 4.8 Distribution of Farms and of Farm Output (Gross Sales) by Economic Class, Canada, 1971, 1976, 1981, and 1986

Gross Sales per Farm	% of Farms				% of Sales			
	1971	1976	1981	1986	1971	1976	1981	1986
$50,000 and over	2.9	12.9	28.5	38.6	29.0	54.8	78.0	86.4
$10,000 – $49,999	28.1	41.9	37.4	32.9	48.6	39.0	19.6	12.0
Under $10,000	69.0	45.4	34.1	28.5	22.4	6.2	2.4	1.6
All Farms	100.0	100.0	100.0	100.0	100.0	100.0	100.0	100.0

Source: Statistics Canada, *Census of Agriculture*, 1971, 1976, 1981, and 1986.

Table 4.9 Distribution of Farms and of Farm Output (Gross Sales) by Economic Class, Canada, 1971, 1976, 1981, and 1986 (Inflation Adjusted)

Gross Sales per Farm (1975 dollars)	% of Farms				% of Sales			
	1971	1976	1981	1986	1971	1976	1981	1986
$50,000 and over (1986 equivalent of $70,300 and over)	9.2	12.9	19.9	29.0	44.5	54.8	68.6	78.4
$10,000 – $49,999 (1986 equivalent of $14,060–$70,299)	44.7	41.7	38.6	36.2	47.3	39.0	27.5	19.0
Under $10,000 (1986 equivalent of under $14,060)	46.1	45.4	41.5	34.8	8.2	6.2	3.9	2.6
All Farms	100.0	100.0	100.0	100.0	100.0	100.0	100.0	100.0

Source: Statistics Canada, *Census of Agriculture*, 1971, 1976, 1981, and 1986, special tabulation.

constant dollar gross sales of $50,000 or more (equivalent of $70,300 or more in 1986) represented 29 percent of the farms and produced 78.4 percent of the total output.

A similar comparison for U.S. data in 1982 constant dollars is provided for 1974, 1978, and 1982 in Figure 4.1. This shows a gradual increase in both farm numbers and sales in constant dollar sales categories of $100,000 or more, while farm numbers and sales in smaller constant dollar gross sales classes have shown a gradual decline. Though of a much smaller magnitude, these rates of change are similar to the changes in Table 4.3 reported in nominal gross sales categories.

Capital Invested

Another measure of farm size is the amount of capital invested. Table 4.10 provides a summary of average capital values per farm by component for Canada and the United States for selected years from 1951 to 1986. Total average capital values per farm in Canada increased from $15,000 in

Figure 4.1 Distribution of Farms and Sales of Agricultural Commodities by Agricultural Sales Class, 1974, 1978, and 1982

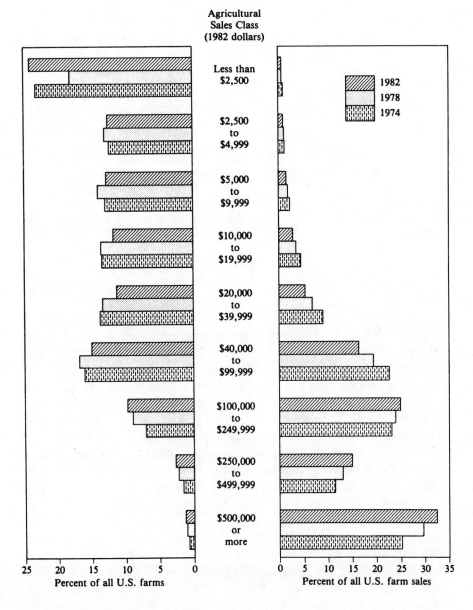

Source: Ahearn, *Financial Well Being of Farm Operators and Their Households*, 1987.

Table 4.10 Average Capital Value per Farm, Canada and the United States, Selected Years ($1,000)

Year	Total		Land and Buildings		Implements and Machinery		Poultry	
	Can.	U.S.	Can.	U.S.	Can.	U.S.	Can.	U.S.
1951	15	25	9	18	3	3	3	4
1961	27	48	18	38	5	6	4	4
1971	65	105	56	83	11	13	8	9
1981	409	415	324	249	55	44	30	22
1986	374	304	273	245	71	38	30	21

Sources: Statistics Canada, *Census of Agriculture*, various years; USDA, *Economic Indicators of the Farm Sector, National Financial Summary, 1988*.

1951 to $409,000 in 1981, before declining from falling land values to $374,000 in 1986. A similar pattern has occurred in the United States, with total capital per farm exceeding Canadian levels through 1981, but falling below Canadian levels by 1986. Roughly 75 to 85 percent of the capital invested in Canada and the United States has been in land and buildings, with average investments per farm in machinery and livestock higher in Canada than the United States in recent years.

Capital values in the European Community are reported in terms of capital formation per year rather than total capital per farm, so direct comparisons with Canada and the United States are not available. Some indication of European capital values may be derived from Table 4.11, however, which summarizes average market values of land and buildings per hectare for Canada, the U.S., and selected European countries over the last two decades. Average land and building values in Canada per *hectare* increased steadily in the 1970s from $247 per hectare in 1971 to a peak of $1520 in 1981, before declining throughout the 1980s to $1077 per hectare in 1988. Land values in the United States have been slightly higher than those in Canada, increasing at a slower rate during the 1970s and bottoming out earlier, by 1987. Land values in Europe are not as readily available, but Table 4.11 shows that they also increased at a rapid rate between 1973 and 1980. In addition, Table 4.11 shows considerably higher land values in Europe than in North America, with values in the more export-oriented countries of France, Denmark, and England ranging around four to seven times those in Canada (i.e., in 1986, 1 ECU = $1.48 Canadian). In the Netherlands and Germany, on the other hand, average land values in recent years have ranged up to twenty times Canadian values. Although European

Table 4.11 Land and Building Values per Hectare, Various Countries, 1971-1988

Year	Canada $ Can.	U.S. $ U.S.	France	Netherlands	Denmark	England	Germany
			—ECU at constant 1984 exchange rates—				
1971	247	501					
1972	267	541					
1973	324	608	1,557		2,114	2,782	
1974	423	746					
1975	536	840					
1976	635	981					
1977	731	1,171					
1978	867	1,312					
1979	1,072	1,552					
1980	1,352	1,821	3,467	12,728	5,300	5,922	15,167
1981	1,520	2,024					
1982	1,517	2,034	3,318		3,819	6,212	16,579
1983	1,448	1,947	3,289		3,910	6,415	17,025
1984	1,379	1,932	3,260		4,419	6,312	17,097
1985	1,278	1,679	3,231	14,822	5,347	6,407	16,471
1986	1,181	1,468	3,202	15,892	6,343	5,503	15,560
1987	1,090	1,349	3,158	14,306	6,484	5,069	14,560
1988	1,077	1,389					

Sources: Canadian data from Statistics Canada, *Value of Lands and Buildings Survey*, various years; data for the U.S. from USDA, *Farm Real Estate: Historical Series Data, 1950-85*, and USDA, *Agricultural Statistics, 1988*; European data from Commission of the European Communities, *The Agricultural Situation in the Community*, various years.

farms tend to be much smaller than Canadian or U.S. farms, the higher landvalues per hectare will generate comparable or higher capital values per farm.

Tenure of Operator

Another important structural dimension of farming is the tenure of the operator. Table 4.12 provides a summary of farms and land area by tenure of operator for Canada and the United States, and the distribution of total hectares owned and rented for Canada and the Euro 10 Economic Community (where other data on tenure are not available). This table shows, for Canada, slow to steady growth in the percent of part owners/part tenants—from 26.2 to 34.0 percent of all farms since 1971—and a corresponding decline—from 68.6 to 59.5 percent—in the proportion of sole owners. Sole tenants show a slow increase from 5.2 to 6.5 percent during this period. On an acreage basis, the percent of land operated by sole owners declined from 44.7 to 32.0 percent from 1971 to 1986, while the proportion operated by part owners/part tenants increased from 49.7 to 59.5 percent. The largest farms were operated by part owners/part tenants. Compared to the overall average size of 572 acres in 1986, part owners/part tenants averaged 1,001 acres. In contrast, full owners averaged only 308 acres and full tenants averaged 742 acres.

Figure 4.2 Index of Relative Change in Land Values (1973 = 100)

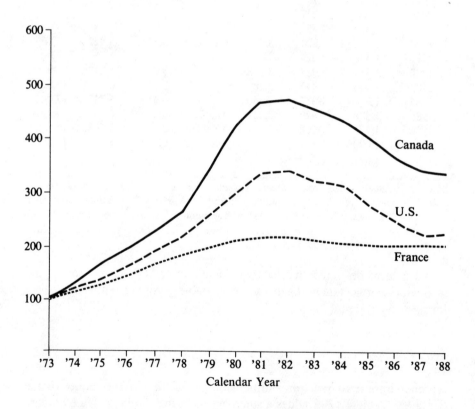

Sources: Canadian data from Statistics Canada, *Farm Land and Buildings Survey*, various years. U.S. data from USDA, *Agricultural Statistics*, 1988. French data from Commission of the European Communities, *The Agricultural Situation in the Community*, various years.

Structural Change in Canadian, United States, and European Agriculture

Table 4.12 Percentage Distribution of Farms and Land Area Classified by Tenure of Operator and Land Arrangement, Canada, U.S., and the Economic Community, Selected Years, 1969-1987

Canada	1971	1981	1986
Percentage of farms operated by			
Full owner	68.6	63.3	59.5
Tenant	5.2	6.2	6.2
Part owner/part tenant	26.2	30.4	34.0
Total hectares farmed by			
Full owner	44.7	39.2	32.0
Tenant	5.6	6.8	8.5
Part owner/part tenant	49.7	54.0	59.5
Total hectares			
Owned	71.9	69.1	63.7
Rented	28.1	30.9	36.3
U.S.	**1969**	**1978**	**1987**
Percentage of farms operated by			
Full owner	62.5	57.5	59.3
Tenant	12.9	12.3	11.5
Part owner/part tenant	24.6	30.2	29.2
Total hectares farmed by			
Full owner	35.3	32.7	32.9
Tenant	13.0	12.0	13.2
Part owner/part tenant	51.8	55.3	53.9
Euro 10 Economic Community	**1970**	**1980**	**1985**
Total hectares			
Owned	63.4	64.3	63.4
Rented	36.4	35.7	36.6

Sources: Canadian data from Statistics Canada, *Census of Agriculture* and special tabulation; U.S. data from USDA, *Agricultural Statistics, 1987*; and U.S. Dept. of Commerce, *Census of Agriculture*, 1987; European data from Eurostat, *Agricultural Statistics Yearbook*, various years.

Data for the United States and Europe show slightly different trends. For the U.S., the percentage of full owners fell slightly to 57.5 percent in 1978, but has since increased to 59.3 percent in 1987. Tenants represent a higher portion of farms in the U.S. than in Canada, and have been showing a slight decrease since 1969. The portion of both farmers who were part owners and part tenants and the share of acreage operated by them peaked in 1978 at 30.2 and 55.3 percent respectively, before declining slightly to 29.2 and 53.9 percent respectively in 1987. For the European Economic Community, the share of total land that is owned versus rented has remained quite constant, at around 63.4 percent over the 1970 to 1985 period.

Form of Business Organization

Organizational structure characteristics are summarized in Table 4.13, which compares only Canada and the United States, as comparable data are unavailable for Europe. The table shows, for Canada, that the individual/family form of sole proprietorship is decreasing in importance as a form of business ownership, from 91.8 percent of all farms in 1971 to 82.3 percent in 1986. Partnerships, on the other hand, have been increasing in importance, from 3.8 percent in 1976 to 11.8 percent in 1986, accounting for 15.1 percent of all capital value at that time. Some increases have also occurred in the number of farms that are family corporations, but the share by nonfamily corporations has remained very low at only 0.4 percent in 1981 and 1986. Since nonfamily corporations control only around 1 percent of all capital in farming, they do not represent a threat to the "family farm" type of operation.

Data for the United States are quite similar, showing a relatively constant but substantial share of farms operated as individual/sole proprietorships over the 1978 to 1987 period. Although agriculture in the United States has been characterized by some of the largest nonfamily corporations in the world, their share of both the total number of farms and acres farmed is very low, and quite similar to the percentages of total capital operated by nonfamily corporations in Canada.

Farm Orientation of the Farm Operator

The farm orientation of the farm operator is reflected in structural statistics by days of off-farm work and commitment to farming as one's primary occupation. Table 4.14 provides a summary of the distribution of farms in Canada and the United States by days of off-farm work, while Table 4.15 summarizes the distribution of Euro 10 farms by their employment status in farming. Table 4.14 shows a much lower percentage of farmers reporting off-farm work in Canada than in the U.S., with the biggest difference occurring in the 200+/229+ day categories. Canada also shows a slightly greater increase in farmers reporting days of off-farm work since 1975, with the greatest increase in the 229+ days category.

Table 4.13 Trends in the Organizational Structure of Agriculture, Canada and the United States, Selected Years, 1971-1987

Canada	1971	1976	1981	1986
		—percent of all farms—		
By Numbers				
Individual/family	91.8	91.5	86.6	82.3
Partnership	5.7	3.8	9.3	11.8
Institutional and other	0.2	0.3	0.3	0.3
Corporation—family	1.9	3.8	3.4	5.2
—other	0.2	0.6	0.4	0.4
All farms	100.0	100.0	100.0	100.0
		—percent of total capital—		
By Capital Value				
Individual/family	84.3	82.4	73.2	68.9
Partnership	9.2	6.1	13.6	15.1
Institutional and other	1.1	0.8	2.0	2.6
Corporation—family	3.7	8.3	9.9	12.1
—other	0.1	1.6	1.2	1.3
All farms	100.0	100.0	100.0	100.0

U.S.		1978	1982	1987
		—percent of all farms—		
By Numbers				
Individual/family		87.1	86.8	86.6
Partnership		10.2	10.0	9.6
Institutional and other		0.4	0.5	0.6
Corporation—family		2.0	2.3	2.9
—other		0.3	0.3	0.3
All farms		100.0	100.0	100.0
		—percent of all acres—		
By Acres				
Individual/family		66.3	65.1	65.0
Partnership		15.6	15.4	15.9
Institutional and other		6.2	6.6	6.7
Corporation—family		10.3	11.4	11.0
—other		1.6	1.5	1.4
All farms		100.0	100.0	100.0

Sources: Canadian data from Statistics Canada, *Census of Agriculture*, Cat. No. 96-701, Table 59 (1971); Cat. No. 96-800, Table 32 (1976); Special Request for 1981 and 1986; U.S. data from U.S. Dept. of Commerce, *1987 Census of Agriculture*.

Table 4.14 Percentage Distribution of Farms by Days of Off-farm Work, Canada and the United States, Selected Years, 1974-1987

Year	All Operators Reporting Any Off-farm Work	Days of Off-farm Work			
Canada		1-24	25-96	97-228	229+
1975	31.1	3.2	6.3	11.0	10.5
1981	38.7	4.1	7.0	13.6	13.9
1986	39.4	3.5	7.4	14.3	14.2
U.S.		1-49	50-99	100-199	200+
1974	54.9	7.3	2.9	8.5	35.7
1982	57.9	7.6	3.3	9.2	37.8
1987	56.9	6.9	3.3	9.1	37.6

Sources: Canadian data from Statistics Canada, *Census of Agriculture*, various years; U.S. data from U.S. Department of Commerce, *Census of Agriculture*, various years.

In the Euro 10 Economic Community, 74.8 percent of all farmers reported no other gainful employment in 1975, decreasing to 69.2 percent in 1985. In 1980 and 1985, 23.7 and 22.9 percent of farmers respectively reported a main source of work other than farming, while 5.1 and 7.5 percent respectively reported having a secondary source of employment. Of those working in farming for the equivalent hours of a full-time nonfarm job, 96.5 and 89.6 percent in 1980 and 1985 respectively reported no other gainful employment. For those working in farming for the equivalent of 50 to 100 percent of a full-time nonfarm job, about three-fourths indicated no other employment, while about 15 percent reported secondary employment. Unfortunately, of those reporting hours of work in farming less than 50 percent of a nonfarm full-time job, about 55 percent also indicated no other employment. This latter statistic identifies a number of underemployed workers in European agriculture. Over all, about 76 percent of all farmers in both 1980 and 1985 reported agriculture as their primary occupation, either with no other gainful employment or with nonfarm work as secondary employment. The corresponding percentage for farming as the primary occupation in the U.S. was around 55 percent for both 1982 and 1987.

Table 4.15 Distribution of Employment in Agriculture, Euro 10 Economic Community, 1975, 1980, and 1985

Category	1975	1980	1985
All Farmers			
Number (1000)	6716[a]	6631	6181
Percentage distribution			
No other gainful employment	74.8[a]	71.2	69.2
Other main employment		23.7	22.9
Other secondary employment		5.1	7.5
Farmers with farm working hours at 100 percent of full time[b]			
Number (1000)		2033	1815
Percentage distribution			
No other gainful employment		96.5	89.6
Other main employment		0.4	0.3
Other secondary employment		3.1	10.1
Farmers with farm working hours at 50–100 percent of full time[b]			
Number (1000)		1081	1036
Percentage distribution			
No other gainful employment		73.9	76.6
Other main employment		8.4	8.1
Other secondary employment		15.8	15.3
Farmers with farm working hours at < 50 percent of full time[b]			
Number (1000)		3514	3228
Percentage distribution			
No other gainful employment		55.2	56.4
Other main employment		41.9	39.9
Other secondary employment		2.9	3.7

[a] Includes an estimate for Greece in 1975.
[b] Farmers working their farms for 100 percent, 50–100 percent, and less than 50 percent respectively of the annual working hours of a full-time worker.
Source: Commission of the European Communities, *The Agricultural Situation in the Community, 1987 Report.*

Canadian Farm Structure Characteristics by Constant Dollar Gross Sales Categories

Additional detail of selected structural attributes by constant 1975 dollar gross sales categories is provided for Canada in Table 4.16. While the number of farms has increased substantially in the larger gross sales categories and declined in the smaller ones, the average constant dollar value of sales in each category has remained stable. While this is not surprising for the categories with fixed levels of gross sales, the unbounded $100,000-plus category has also been quite constant (with only small increases).

The number of farms reporting hired labor in each gross sales category has changed roughly proportional to the change in farm numbers per category, with the exception of the smallest categories, where the number of farms reporting hired labor has increased since 1975 despite declines in farm numbers. The average weeks of hired labor noted per farm reporting hired labor declined significantly since 1971 for the largest farms with $100,000 or more gross sales. It decreased moderately for the $50,000-$99,999 sales category, showed little change for the middle categories, but increased for the smallest. By 1986, however, the largest farms with $100,000 or more gross sales still reported the greatest level of hired labor among size categories, with an average of 104.3 weeks per farm.

Off-farm work was reported by about 14 percent of the largest farms with $100,000 or more gross sales in 1986, which was about the same percentage as reported for farms of this size category in 1971. This percentage increased steadily across farms in smaller gross sales categories, reaching about 60 percent of the smallest farms with constant 1975 dollar sales of $2,500-$4,999, and under $2,500 in 1986. The average days reported per farm reporting off-farm work has increased since 1971 for all gross sales categories, except those under $2,500.

Table 4.16 also shows that the exclusive owner operator form of land tenure has been declining—from 68.6 percent in 1971 to 59.5 percent in 1986. The sole tenant form of land tenure has not changed much, increasing only slightly from 5.2 percent in 1971 to 6.5 percent in 1986. Joint owner and tenant operations increased steadily, however, growing from 26.2 percent in 1971 to 34.0 percent in 1986. The lowest percentage of sole ownership has been found among the largest operators; this percentage is also declining the fastest among this group. In a similar fashion, the percentage of farms with a joint owner-tenant has been both the highest and the fastest growing on the largest farms.

Table 4.16 Selected Farm Data Classified by Indexed Sales Class, Canada, Selected Years

Item & Year	Total Farms	Value of Gross Farm Sales Expressed in Constant 1975 Dollars							
		$100,000 and over	$50,000–$99,999	25,000–$49,999	$10,000–$24,999	$5,000–$9,999	$2,500–$4,999	Under $2,500
Total Number of Farms								
1971	365,334	8,603	24,987	58,975	104,263	60,363	37,427	70,716
1976	337,782	12,349	31,309	59,308	81,492	45,789	37,871	69,664
1981	317,758	22,557	40,811	57,706	64,806	36,849	30,822	64,207
1986	293,089	35,042	50,081	51,687	54,460	32,112	25,806	43,901
Average Indexed $ per Farm								
1970	21,791	217,049	67,176	34,902	16,477	7,399	3,719	891
1975	26,253	221,165	68,053	35,488	16,663	7,348	3,683	1,000
1980	36,242	225,280	68,929	35,994	16,849	7,297	3,646	1,108
1985	50,439	229,432	70,950	36,644	16,789	7,326	3,690	1,167

continued . . . .

Table 4.16 continued

Farms Reporting Hired Agricultural Labor								
1970	129,062	7,339	17,922	31,854	39,007	16,099	8,221	8,620
1975	98,411	9,525	18,533	24,684	22,672	9,497	6,621	6,879
1980	114,858	17,471	25,052	25,998	20,940	9,609	6,863	8,925
1985	141,817	30,170	36,136	28,213	21,784	10,153	6,851	8,376
Average Weeks Hired Labor by Farms Reporting								
1970	28.9	183.0	44.9	22.2	14.7	9.7	8.6	9.1
1975	36.4	167.4	40.6	23.5	17.1	14.5	12.6	11.8
1980	38.9	124.1	38.8	24.1	17.6	14.7	13.5	11.8
1985	40.3	104.3	33.4	22.9	17.0	13.6	11.3	15.2
Farms Reporting Off-Farm Work								
1970	129,239	1,279	4,360	12,649	30,216	23,725	18,165	38,845
1975	114,546	1,396	3,818	9,792	22,515	19,282	19,326	38,417
1980	123,071	3,592	7,643	14,243	24,080	18,508	17,320	37,694
1985	115,608	4,950	10,061	15,678	24,531	18,236	15,530	26,591

continued . . .

Table 4.16 continued

Average Days by Operators Reporting								
1970	154.2	98.0	85.3	94.4	121.1	152.3	174.3	200.9
1975	172.1	108.0	93.2	109.8	142.0	172.6	189.9	206.6
1980	170.9	116.9	101.2	117.6	153.0	181.6	195.7	205.2
1985	172.9	115.4	111.0	135.5	168.5	190.0	198.4	206.7
Operators Classified by Tenure								
% Operator owner								
1971	68.6	49.2	48.9	55.0	64.2	73.0	79.3	86.3
1976	66.1	42.9	45.3	53.6	62.5	70.4	76.7	86.0
1989	63.4	42.5	44.5	52.3	61.9	70.0	76.6	84.1
1986	59.5	38.8	43.5	50.9	61.9	71.1	77.1	82.5
% Tenant								
1971	5.2	3.5	4.1	3.2	4.5	6.0	6.8	7.0
1976	5.7	4.3	3.8	3.8	5.9	7.6	7.0	6.1
1981	6.2	3.9	3.6	4.8	7.1	9.0	7.7	6.5
1986	6.5	3.2	3.8	5.7	8.2	8.7	8.2	8.6
% Part owner/part tenant								
1971	26.2	47.3	47.0	41.8	31.3	21.0	13.9	6.7
1976	28.2	52.8	50.9	42.6	31.6	22.0	16.3	7.9
1981	30.4	53.6	51.9	42.9	31.0	21.0	15.7	9.4
1986	34.0	58.0	52.7	43.4	29.9	20.2	14.7	8.9

Source: Statistics Canada, *Census of Agriculture*, 1971, 1976, 1981, 1986, special tabulation.

Macro Structure Characteristics

In addition to structural characteristics at the farm level, other characteristics may also be important at the sector level. This section will examine changes in productivity, relative prices, and self-sufficiency ratios.

Productivity

Productivity represents the conversion of inputs into outputs, and may be measured many ways. Productivity is often measured as yields per area planted or output per person hour. These are partial productivity measures, however, and must be carefully utilized, since improvements may be accomplished with tremendous increases in other inputs, such as fertilizer and machinery, which add to costs of production. Between 1970 and 1985, yields of wheat in bushels per acre increased at an average annual rate of only 0.1 bushels per acre for Canada. This compares to 0.2 bushels per acre for Australia, 0.5 bushels per acre for Argentina and the United States, and 2.1 bushels per acre for France (USDA, *Agricultural Statistics*, 1988). Some of the differences between wheat productivity in North America and the European Community can be explained by the increased use of chemicals (intensive cereals management systems) in response to higher producer prices; but some of the low increases in Canada can be explained by institutional rigidities in grading and marketing of grain, and in the registration restrictions on new varieties.

An alternative to single factor measures of productivity (like yields per acre) is called multifactor productivity, which measures changes in output relative to all inputs. While this measurement provides a more comprehensive assessment, it is unfortunately not widely available for countries other than Canada and the U.S. Figure 4.3 provides a comparison of these two countries, since 1970, for the entire agricultural sector. Data are only available for Canada through 1984, but the figure indicates that overall multifactor productivity in the two countries has increased at roughly the same rate. Since 1977, however, it would seem that the U.S. has had a slight edge, particularly considering the large increases there from 1983 to 1985, and the likely impact in Canada from droughts on the prairies in the late 1980s.

Figure 4.3 Index of Multifactor Productivity Change, Canada and the U.S. (1977 = 100)

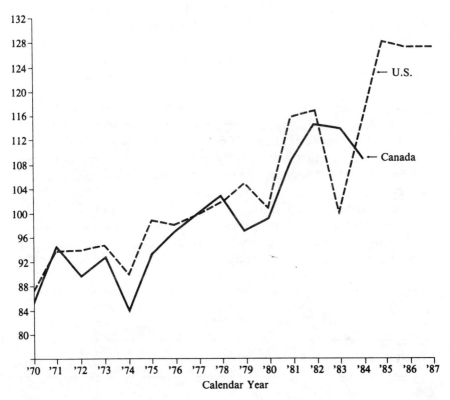

Source: Canadian data from Brinkman and Prentice, *Multifactor Productivity in Canadian Agriculutre*, 1983, and updated with unpublished Agriculture Canada data. U.S. data from USDA, *Economic Indicators of the Farm Sector*, 1989.

Ratios of Prices Received to Prices Paid

Table 4.17 provides a summary of price indices for prices received and paid by farmers, revealing trends in agricultural terms of trade in agriculture within the various countries. Farmers in Canada and the U.S. have seen prices paid go up considerably faster than prices received, while these price relationships have been maintained at a constant ratio for the European Community. These ratios, in part, reflect the impact of high European support prices and price advantages received by European farmers. It should be pointed out, however, that North American farmers have also received substantial benefits through deficiency payments, which are not reflected in these price ratios. Furthermore, because of increasing productivity in input use, it is not necessary to maintain equal price levels for prices received and paid to maintain agricultural prosperity.

Self-Sufficiency Ratios

Self-sufficiency ratios reflect the relative capacity for a country to provide sufficient production to meet domestic consumption requirements. Canada has traditionally been an exporter of grain, increasing its self-sufficiency in cereals from 130 percent in 1970 to 200 percent in 1985, before declining to 145 percent in 1988 because of the North American drought (see Table 4.18). Substantial increases in cereal self-sufficiency and export capacity were also experienced by the U.S. and France. Germany and the United Kingdom also changed from a position of net cereal imports to basically a position of self-sufficiency by 1988. Overall, the European Community had achieved 121 percent self-sufficiency in cereals by 1985, along with

Table 4.17 Indices of Prices Paid and Received by Farmers (1975 = 100), Canada, U.S., and the European Community, Selected Years, 1975–1988

Country and Index	1975	1980	1985	1988
Canada				
Prices received	100	136	140	133
Prices paid	100	158	192	198
Ratio prices received to paid	1.00	.86	.73	.67
U.S.				
Prices received	100	133	127	126
Prices paid	100	155	183	182
Ratio prices received to paid	1.00	.86	.69	.69
European Community (Euro 10)				
Prices received	100	149	214	218
Prices paid	100	150	214	206
Ratio prices received to paid	1.00	.99	1.00	1.06

Sources: Canadian data from Statistics Canada, *Farm Input Price Index*, various years; U.S. data from USDA, *Agricultural Statistics*, various years; data for European community from Eurostat, *Agricultural Statistical Yearbook*, 1989.

100+ percent self-sufficiency in meat. These latter changes have been reflected first in reduced imports into the European Community, followed by subsidized exports to reduce surpluses.

Summary of Structural Conditions and Implications

An overall summary of current structural conditions for Canada is provided in Figure 4.4, which compares the myth and reality of modern-day agriculture in both Canada and the United States. Many of these conditions also hold for Europe, although European agriculture is dominated by smaller holdings, extremely high land values, operators dependent primarily on agriculture for their livelihood, and high levels of government support.

In Canada and the U.S., the approach to structural change has been to promote competitiveness based on efficiency, whereby individual farmers can adjust the size and organization of their farms through the purchase of reasonably priced land. In parts of Europe, on the other hand, structural policy has focused on attempts to maintain farmers on the land through a variety of more directly aimed structural policies and support programs. While the Europeans have been successful in maintaining a higher proportion of smaller farms than found in the U.S. or Canada, these policies have also created exceptionally high land values, which, in turn, have prevented further restructuring from taking place through the market. Furthermore, the combination of high European land values, relatively small holdings, and high dependence on agriculture by farmers, has created continuing pressures for the maintenance of high levels of support. Each country may choose its own structural policy, but these policies may generate substantially different costs for implementation.

Table 4.18 Trends in Self-sufficiency Percentages, Selected Countries, 1970–1988

	1970	1975	1980	1985	1988
Canada					
Cereals	130	175	177	200	145
Food	131	134	141	151	
U.S.					
Cereals	113	160	157	172	94
Food	115	134	131	138	
Euro 10					
Cereals	91	92	101	121	
Meat	91	97	100	102	
Milk	100	100	101	101	
France					
Cereals	134	145	169	207	209
Food	112	113	122	126	
Germany					
Cereals	71	80	90	95	104
Food	83	84	86	92	
United Kingdom					
Cereals	59	65	95	112	113
Food	70	72	81	87	

Sources: Individual countries from USDA, *World Agricultural Trends and Indicators, 1970-1988*, and Euro 10 data from Commission of the European Communities, *The Agricultural Situation in the Community*, various years.

Figure 4.4 Farming: Myth and Reality

Paradigm	Reality
An important source of employment	A minority occupation
Homogeneous structure and dispersed resource ownership	Heterogeneous structure, with a high degree of concentration
Predominantly full-time businesses with most of income from farming	Majority are part-time farmers, with off-farm income largest single income source
Resource based, biological activity	Capital and technology intensive industry with some commodity production systems similar to manufacturing
Farm labor provided primarily by the farm family	Hired labor input stable and increasing relative to diminishing family labor input
Entrepreneurial freedom in production and marketing decisions	Decisions increasingly made by PDR sector or constrained by marketing boards and governments
Arms length transactions in competitive and decentralized markets	More closed pricing systems in highly concentrated markets
Ease of access	Declining opportunities and higher barriers to entry
Resilience in economic adversity	Increasingly vulnerable to economic instabilities
Integration of resource ownership and operation, with full ownership by sole proprietors	Renting more of resources used and increasing incidence of corporations, partnerships, and other forms of multiple ownership and joint operation
Farmers the dominant element in rural society	Farmers a minority in rural communities
Social and cultural distinctiveness of farm people	Increasingly indistinguishable from nonfarm people

Notes

1. For example, Canada and the United States share a 4000-mile common border and therefore face similar climatic and land conditions over an extensive area. The initial development of agriculture in both countries was undertaken largely by European immigrants through the clearing and homesteading of previously unfarmed land, in a political environment supportive of private ownership. Both countries have highly industrialized nonfarm economies in which many of the same firms provide inputs in both countries. Traditionally, technology has flowed freely between the two countries; farm machinery is imported into Canada duty free; and there is relatively free bi-country trade in a number of important commodities (beef, pork, feed grains). World market forces also create similar impacts in a North American context for internationally traded products and, to a lesser extent, land prices and energy.

References

Ahearn, M. *Financial Well Being of Farm Operators and Their Households.* Washington, D.C.: ERS, USDA, ERS No. 563, 1986.

Brinkman, G.L. *Structural Change in Canadian Agriculture in the 1980s.* Ottawa: Farm Development Policy Directorate, Policy Branch, Agriculture Canada, 1989.

Brinkman, G.L., and B.E. Prentice. *Multifactor Productivity in Canadian Agriculture: An Analysis of Methodology and Performance, 1961-1980.* Ottawa: Regional Development Branch, Production Development Policy Directorate, Agriculture Canada, 1983.

Brinkman, G.L. and T.K. Warley. *Structural Change in Canadian Agriculture: A Perspective.* Ottawa: Regional Development Branch, Production Development Policy Directorate, Agriculture Canada, 1983.

Commission of the European Communities. *The Agricultural Situation in the Community.* Luxembourg: Office for Official Publications of the European Communities, various years.

Eurostat. *Agriculture Statistics Yearbook.* Theme 5, Series A. Luxembourg: Office for Official Publications of the European Communities, various years.

Shertz, L. et al. *Another Revolution in U.S. Agriculture.* Washington, D.C.: USDA, ESCS-AER441, 1979.

Statistics Canada. *Census of Canada, Agriculture.* Ottawa: Ministry of Supply and Services, 1971, 1976, 1981, 1986.

———. *Farm Input Price Index.* Catalogue No. 62-004. Ottawa: Ministry of Supply and Services, various years.

———. *Farm Land and Buildings Survey.* Ottawa: Ministry of Supply and Services, various years.

United States Department of Agriculture. *Agricultural Statistics, 1988.* Washington, D.C.: United States Government Printing Office, 1988.

———. *A Time to Choose: Summary Report on the Structure of Agriculture.* Washington, D.C.: United States Government Printing Office, 1981.

———. *Conceptual Framework, Overview and Summaries of Contributing Projects to U.S. Competitiveness in the World Wheat Market: A Prototype Study.* Washington, D.C.: USDA, ERS, 1986.

———. *Economic Indicators of the Farm Sector, Production and Statistics, 1987.* ECIFS 7-5. Washington, D.C.: USDA, ERS, 1989.

———. *Economic Indicators of the Farm Sector, National Financial Summary, 1988.* ECIFS 8-1. Washington, D.C.: USDA, ERS, 1989.

———. *Farm Real Estate: Historical Series Data, 1950-85.* Statistical Bulletin No. 738. Washington, D.C.: USDA, ERS, 1985.

———. *Structure Issues of American Agriculture.* ESCS-AER438. Washington, D.C.: USDA, ERS, 1979.

———. *World Agricultural Trends and Indicators.* Statistical Bulletin No. 781. Washington, D.C.: USDA, ERS, 1989.

U.S. Department of Commerce. *Census of Agriculture.* Washington, D.C.: Bureau of the Census, various years.

———. *United States Summary and State Data.* Volume 1, Part 51. Washington, D.C.: U.S. Bureau of the Census, 1974, 1978, 1982, 1987.

5

The Crisis in European and North American Agriculture

Michele M. Veeman and Terrence S. Veeman

Introduction

Agriculture in Western Europe, the United States, and Canada exhibits similar patterns of structural adjustment, which arise from the common experience of relatively high income developed nations. Despite considerable differences in the size of these economies, in the nature of their agricultural sectors, and in the nature and orientation of many government programs for agriculture, each of these regions exhibits declining numbers of farms and farmers, declining disparity of farm family income from incomes of all families (with the probable exception of Southern Europe, where underemployment in agriculture remains a serious problem), and faster growth in the national economy than in the agricultural sector. Some of these features are shown in Table 5.1. From an economic perspective, these are desired changes that are part and parcel of the process of economic growth. From a social perspective, these changes are often costly in human terms. The costs of adjustment have political implications that depend on the extent and distribution of the social and economic effects. These various influences have led to differing levels and types of market intervention and support for agriculture in the three regions.

Currently, however, there is considerable dissatisfaction that support levels for agriculture in Europe and North America have reached untenable levels, that domestic policies for agriculture in our respective nations are sorely in need of reform, and that the improvement of international trading arrangements for agriculture is being severely impaired. Efforts to reform domestic agricultural and international trade policy will not be easy, and will need to recognize the combination of economic, social, and political

Table 5.1 Key Indicators for Selected Developed Nations and Their Agricultural Sectors

Item	Canada	USA	U.K.	France	Germany	Japan
Population (millions), 1988	26.1	246.0	57.1	55.8	61.0	122.6
GNP per capita (US$), 1987	15,160	18,530	10,420	12,790	14,440	15,760
Annual growth rate of GNP per capita (%), 1965-87	2.7	1.5	1.7	2.7	2.5	4.2
Agriculture's share in GDP (%), 1965	6	3	3	8	4	9
Agriculture's share in GDP (%), 1987	3	2	2	4	2	3
Growth rate in ag. production (%/year), 1965-80	0.7	1.0	1.6	1.0	1.4	0.8
Growth rate in ag. production (%/year), 1980-87	2.6	3.5	3.2	2.6	1.9	0.8
Value-added in agriculture (million US$), 1987	10,449	87,482	8,567	26,979	16,541	65,384
Employment in agriculture ('000), 1987	475	3,373	592	1,489	1,327	4,906
Share of labor force in agriculture (%), 1965	10	5	3	18	11	26
Share of labor force in agriculture (%), 1987	4.0	3.0	2.4	7.1	5.2	8.3
Arable and permanent cropland (million ha.), 1986	46.0	189.9	7.0	19.0	7.5	4.7
Agricultural labor force/1000 ha., 1985	9	10	64	87	117	1,074
Fertilizer use (kg./ha.), 1986	47	92	380	309	428	427
Wheat yield (kg./ha.), 1987 or 1988	1,931	2,533	6,137	5,967	6,838	3,670
Value of agricultural exports (billion US$), 1987	7.3	31.3	10.0	23.6	15.1	0.9
Value of agricultural imports (billion US$), 1987	5.4	22.0	18.3	17.7	29.7	21.0
Disposable income spent on food (%), 1985	10.7	10.0	13.0	15.8	19.1	–

Sources: World Bank, *World Development Report* (1988, 1989); US Department of Agriculture, *World Agricultural Trends and Indicators 1970-88* (June 1989); and Commission of the European Community, *The Agricultural Situation in the Community: 1988 Report* (1989).

influences that have molded agricultural policies in each region. Nonetheless, with suitable compensation to those disadvantaged by change, international and national welfare could be enhanced by improved policies for agriculture in rich nations.

Initially, this chapter provides an overview of the nature of the crisis facing European and North American agriculture, outlining its economic and structural roots. It briefly summarizes the major market and price policies for agriculture that have arisen in the European Community (EC), the United States, and Canada. It then outlines the pressures for reform of domestic agricultural policies and international agricultural trade, and discusses some major obstacles to achieving these ends. In attempting to find a possible solution to the current economic and political impasse, this chapter then examines the general set of policies that seem appropriate for rich-nation agriculture. This inevitably involves the authors' personal perception of the appropriate role of government in agriculture and the respective circumstances in which market forces do, and do not, work in the agricultural sector. It is concluded that the agricultural sectors of Europe and North America (and certainly Japan) must be increasingly subjected to market pressures, particularly in the realm of output and resource allocation decisions, prices, and trade patterns. In these areas, more market discipline is needed and much less government intervention, protection, and subsidy. The proper role for government in agriculture, it is suggested, is to deal with income distribution, risks faced by farmers, questions of environment, and agricultural and food safety.

The Nature of the Crisis and Its Roots

The 1980s have proved to be a most challenging decade for farm families, rural communities, and many of the industries linked to agriculture. In the first half of this decade, global recession contributed to lower volumes of trade and declining levels of world prices for farm products, measured in real as well as nominal terms. OECD (1988, p. 188) reports that real agricultural commodity prices (measured in constant U.S. dollars) declined 32 percent from 1980 to 1986; by 1986/87 these were at their lowest level since 1948. Farmers in most high-income countries have been sheltered from the full effects of declining world prices for farm products. Depressed prices, stemming initially from depressed demand, and reinforced by the effects of increased levels of subsidy and support in many countries that encouraged maintenance of production levels, have been most evident for farm sectors dependent on and linked, through trade,

Table 5.2 Economic Trends in Western European Agriculture, 1977 to 1989 (Index Numbers: Average 1979-81=100)

Year	Final Production (value)	Productivity of Material Inputs[2]	Deflated Farmgate Prices (real)	Deflated Input Prices (real)	Ratio of Farmgate to Input Prices	Net Value-Added in Agriculture (real)	Farm Labor	Net Value-Added Per Person (real)	Nominal Amount (bill ECU)	Nominal Amount (bill US$)	% Share of EC Budget	% Share of GDP
1977	90.3	98.6			106.3	103.8	107.1	102.1	6.8	7.8		0.48
1978	95.0	99.0			107.8	104.7	104.6	105.4	8.7	11.1	76.3	0.55
1979	98.6	98.2			104.5	104.6	102.3	102.9	10.4	14.3	72.7	0.59
1980	100.5	99.7			99.2	96.2	100.5	96.3	11.3	15.8	69.5	0.56
1981	100.9	102.2	98.8	101.5	97.1	99.2	97.3	100.8	11.1	12.4	61.4	0.50
1982	105.6	104.8	98.6	99.8	97.3	109.6	94.1	111.5	12.4	12.2	59.9	0.51
1983	105.5	103.5	95.9	99.4	95.4	106.5	93.9	105.0	15.9	14.2	63.7	0.61
1984	108.4	106.2	92.9	99.2	93.4	108.1	92.2	108.3	18.4	14.5	67.4	0.66
1985	107.9	104.3	89.0	95.0	93.8	102.8	90.5	103.0	19.8	15.1	70.3	0.66
1986	110.1	105.5	86.0	89.5	96.9	105.2	89.3	104.9	22.2	21.8	62.9	0.63
1987	108.8	104.2	81.7	83.2	98.2	98.9	87.0	99.8	27.7	32.0	77.5	0.75
1988			79.8E	82.5E					26.4	31.2	65.7F	0.75F
1989									26.7	30.1	62.9F	0.68F

1 Final production appears to be analytically equivalent to gross farm receipts. Gross value-added equals the value of final production less the cost of intermediate consumption inputs. Net value-added equals gross value-added minus depreciation. Net value-added per person is calculated in terms of per annual work unit (full-time labor equivalents).

2 The input basket used for aggregate input calculations is narrower in EC than in North American statistics and consists of only intermediate consumption or material inputs.

3 The European Community is composed of the original six founding members (Belgium, Luxembourg, France, Italy, West Germany, and the Netherlands) plus Denmark, Ireland, and the United Kingdom (joining in 1973), Greece (1981), and Spain and Portugal (1986).

Source: Commission of the European Communities, *The Agricultural Situation in the Community*, various annual reports.

to world markets. Meanwhile, pressures of higher costs, arising largely from unprecedented levels of real interest rates, affected all sectors of agriculture as national monetary authorities, fearing resurgence of the inflationary pressures of the 1970s, reacted to oil price shocks in the late 1970s.

Farmers in the United States, Canada, and Western Europe, as in many other regions of the world, have experienced, to varying degree, declining terms of trade, pressures on net farm incomes, and falling asset values, relative to the buoyant economic conditions that prevailed for much of the farm sector during the 1970s. Some of these impacts are illustrated in Tables 2 to 4. Adverse effects on agriculture have been less pronounced for farmers in Western Europe and those in the more highly protected of the North American farm sectors. The impacts of depressed world markets for farm products, coupled with higher levels of real interest rates, have been disastrous for the group of heavily indebted developing countries. The effects on their balance of payments, debt servicing costs, and abilities to import, further constrain the levels of world trade and prices for farm products.

World prices for agricultural products have tended to increase somewhat (in nominal terms) since 1987, but economic problems have persisted into the last half of the 1980s for some segments of the industry. In particular, the "export subsidy war" continues to have an adverse effect on world grain prices, despite a temporary increase in grain prices due to the 1988 drought in North America.

Some features of the economic problems of agriculture in the 1980s arose from the rather different economic pressures of the 1970s. In that earlier decade, world trade and prices of farm goods rose, farm incomes increased, and farm asset values escalated in North America, as well as in other regions. Agriculture in the Economic Community was considerably shielded from these pressures by the common agricultural policy. Sparked by the commodity price boom of the 1970s, combined with the pressures of inflation in the wider economy, and the effect of low interest rates, the value of farmland in North America rose dramatically in the 1970s, reaching levels that would yield only low rates of return to owner-operators' labor, but based on the expectation of continued capital appreciation. The consequent increased levels of farm debt appear to have been concentrated within a relatively small proportion of all farmers, giving this group of highly leveraged and highly indebted farmers severe cash-flow problems as interest rates climbed and farm prices fell in the early to mid-1980s. Farmland values declined in the 1980s in North America and parts of

Table 5.3 Economic Trends Relating to Agriculture, United States, 1973 to 1989

Year	Gross Farm Income	Total Prod. Expenses	Net Farm Income Current Dollars	Net Farm Income Deflated 1982 $	Off Farm Income	Total Farm Assets	Farm Land Values	Total Farm Debt	Debt/Asset Ratio
			Billion Dollars				Billion Dollars		
1973	98.9	64.6	34.4	69.4					
1974	98.2	71.0	27.3	50.5					
1975	100.6	75.0	25.5	43.1					
1976	102.9	82.7	20.2	32.0		590.4	453.5	97.0	16.4
1977	108.8	88.9	19.9	29.5		651.6	509.1	110.9	17.0
1978	128.4	103.2	25.2	34.9		777.2	601.9	127.4	16.4
1979	150.7	123.3	27.4	34.9		907.8	706.2	151.6	16.7
1980	149.3	133.1	16.1	18.8	34.7	995.6	782.4	166.8	16.8
1981	166.4	139.4	26.9	28.6	35.8	996.7	784.7	182.3	18.3
1982	163.5	140.0	23.5	23.5	36.4	961.0	748.8	189.5	19.7
1983	153.1	140.4	12.7	12.2	37.0	944.3	738.7	192.7	20.4
1984	174.9	142.7	32.2	29.9	38.9	846.7	637.7	190.7	22.5
1985	166.4	134.0	32.4	29.2	42.6	746.4	555.9	175.1	23.5
1986	160.4	122.4	38.0	33.4	44.6	689.5	507.3	155.1	22.5
1987	171.6	128.0	43.6	37.2	46.8	764.9	577.0	143.1	18.7
1988	177.6	135.0	42.7	35.2	51.7	810.4	607.9	138.4	17.1
1989F	190.0	141.0	48.0	38.0	54.0	849.0	648.0	136.0	16.0

continued . . .

Table 5.3 continued

Year	Prices Received by Farmers	Input Prices Paid	Terms of Trade	Farm Output Index	Farm Input Index	Total Factor Prod'y	Average Land Value /Acre
			Index Numbers: 1977 = 100				
1973	98	72	136	93	98	95	53
1974	105	81	130	88	98	90	66
1975	101	89	113	95	97	99	75
1976	102	95	107	97	98	98	86
1977	100	100	100	100	100	100	100
1978	115	109	106	104	102	101	109
1979	132	125	106	111	105	105	125
1980	134	139	96	104	103	101	145
1981	139	151	92	118	102	116	158
1982	133	159	84	116	99	117	157
1983	135	159	85	96	97	99	148
1984	142	162	88	112	95	119	146
1985	128	157	82	118	92	128	128
1986	123	150	82	111	87	127	112
1987	126	151	83	110	86	128	103
1988	138	160	86	101	85	120	106
1989F				109			

Sources: U.S. Department of Agriculture, *Agricultural Statistics 1988* (Washington: 1988), and USDA-ERS, "Agricultural Outlook" (January/February 1990).

Western Europe, as well as in other regions of the western developed world, particularly New Zealand and Australia.

Economic pressures on agriculture prior to and during the 1980s reflect the increased protectionism and support of domestic agriculture that has occurred in virtually all high-income countries. Successful economic development is inseparable from declining numbers of farms and farmers, but farm organizations and associated agribusiness sectors have been effective lobbyists for support and protection of less competitive, higher cost farm sectors. The use of market intervention systems in the pursuit of higher and more stable farm prices for domestic agriculture in most high-income countries has depressed world prices and contributed to the considerable variability in residual world markets for most farm products. This is reflected in Table 5.5, which measures producer subsidy equivalents for the United States, Canada, and the European Community, over the early and mid-1980s.

Table 5.4 Economic Trends in Canadian Agriculture, 1971 to 1990

Year	Total Cash Receipts bill. C$	Net Farm Income bill. C$	Real[1] Net Farm Income 1981 $	Farm Product Price Index 1981=100	Farm Input Price Index 1981=100	Labor Force Employed '000	Current Farm Land[2] Values bill. C$	Real[1] Value of Land per Acre 1981=100	Real Interest Rate[3] %	Farm Debt bill. C$
1971	4.7	1.41	3.34	34.0		510	16.9	38.6	3.58	4.6
1975	10.2	3.98	6.80	69.2		483	36.6	60.3	-1.99	7.8
1976	10.1	3.11	4.94	66.3	59.1	472	43.6	66.5	2.58	9.1
1977	10.2	2.53	3.73	65.1	61.6	464	50.0	70.8	0.16	10.4
1978	12.1	3.08	4.17	74.9	68.8	474	59.3	77.3	0.89	12.1
1979	14.4	3.28	4.06	86.9	80.3	484	73.2	87.5	3.71	14.2
1980	16.0	2.74	3.08	94.0	88.0	479	92.0	100.0	4.06	15.9
1981	18.5	4.01	4.01	100.0	100.0	488	103.3	100.0	6.79	18.3
1982	18.9	2.89	2.61	98.1	103.1	465	103.1	90.1	5.07	20.0
1983	18.8	2.07	1.77	97.6	103.9	480	98.4	81.3	5.37	21.0
1984	20.4	2.66	2.17	102.9	106.5	480	93.6	74.2	7.71	21.7
1985	19.7	3.38	2.66	97.2	106.5	475	86.6	66.1	6.63	22.5
1986	20.4	4.47	3.38	93.1	108.4	467	80.1	58.7	6.43	23.6
1987	21.1	4.15	3.00	92.4	110.1	461	73.9	51.9	5.12	23.1
1988	22.1	3.46	2.41	96.8	113.2	444	73.0	49.3	6.78	22.7
1989E	22.7	4.92	3.25	100.2	116.1	429			8.06	
1990E	20.9	2.81	1.77							

1 Nominal or current value deflated by respective Consumer Price Index (CPI) number.
2 The land series includes both land and buildings.
3 Canadian bank prime lending rate minus the rate of inflation. See Agriculture Canada, *Market Commentary: Farm Inputs and Finance* (December 1989): 79.

Sources: Based on data in Statistics Canada, *Agriculture Economic Statistics, Catalogue No. 21-603E*, and Agriculture Canada, *Market Commentary* (December 1989).

Table 5.5 Producer and Consumer Subsidy Equivalents by Nation, 1984 to 1988 (net PSE and CSE for all agricultural products as a percent of producer gross revenues and consumer expenditures, respectively)

		1984	1985	1986	1987(E)	1988(P)
Australia	PSE	10	14	16	11	10
	CSE	-8	-11	-12	-10	-6
Austria	PSE	33	39	50	53	48
	CSE	-36	-43	-55	-60	-51
Canada	PSE	33	39	49	46	43
	CSE	-28	-34	-41	-38	-31
EEC (10/12)	PSE	33	43	52	51	46
	CSE	-28	-40	-52	-51	-42
Finland	PSE	60	67	70	71	70
	CSE	-59	-67	-73	-74	-70
Japan	PSE	67	69	76	77	74
	CSE	-41	-45	-53	-54	-53
New Zealand	PSE	18	23	33	14	8
	CSE	-9	-12	-8	-8	-6
Sweden	PSE	38	40	54	61	58
	CSE	-31	-41	-56	-63	-58
United States	PSE	28	32	43	41	34
	CSE	-20	-21	-26	-23	-16
Average	PSE	34	41	51	50	45
	CSE	-27	-33	-44	-43	-37

Note: The producer subsidy is the payment that would be required to compensate producers for the loss of income resulting from the removal of a given policy measure. It is a partial equilibrium measure, determining output at world prices, but not allowing for tax and subsidies in other sectors. The consumer subsidy equivalent measures the extent to which the transfers to producers are provided by consumers through higher prices. E denotes estimate, P denotes provisional.

Source: OECD, *Agricultural Policies, Markets and Trade: Monitoring and Outlook 1989* (July 1989).

As is evident in Table 5.6, the extent of transfers from taxpayers and consumers to farm producers in Europe and North America has grown to unprecedented and extremely costly levels. Government transfers, via taxpayers, have been approximately $100 billion in each of 1986, 1987, and 1988 for the United States, the EC-12, and Canada combined. Transfers from consumers in these nations in aggregate have been slightly higher than those from taxpayers in each of these years. Total taxpayer and consumer transfers to farm producers in 1988 averaged US$33,500 per farm in the United States (on the basis of 2.2 million farms), US$26,600 per farm in Canada (across 293 thousand farms), and US$13,600 per farm in Western

Table 5.6 Total Transfers Associated with Agricultural Policies in Developed Nations

Country	Transfers from Taxpayers			Transfers from Consumers			Total Transfers[2]		
	1983-85 Average	1986	(E)[1] 1987	(P) 1988	1983-85 Average	1986	(E) 1987	(P) 1988	1983-85 Average	1986	(E) 1987	(P) 1988
	(Billions of US$)									
United States	47.1	62.1	56.3	51.3	24.2	31.1	30.6	23.4	69.7	91.9	86.1	73.8
Canada	3.2	4.3	5.2	4.6	2.8	3.5	3.5	3.4	5.9	7.7	8.6	7.8
Australia	0.6	0.7	0.6	0.7	0.4	0.4	0.4	0.3	0.9	1.1	1.0	1.0
New Zealand	0.3	0.9	0.1	0.2	0.1	0.1	0.1	0.1	0.4	0.9	0.1	0.2
Japan	10.8	13.8	17.5	17.2	29.2	49.0	55.9	59.8	34.7	54.1	62.6	63.8
EEC-10	22.0	31.0	n.a.	n.a.	36.7	67.4	n.a.	n.a.	58.1	97.7	n.a.	n.a.
EEC-12	n.a.	31.7	38.0	45.0	n.a.	77.6	86.6	75.3	n.a.	108.7	123.9	119.4

1 E: Estimate; P: Provisional; n.a.: not applicable.
2 Total transfers equal transfers from both taxpayers and consumers less any budget revenue earned by government. Budget revenues are minimal in all nations except Japan.

Source: OECD, *Agricultural Policies, Markets and Trade: Monitoring and Outlook 1989* (July 1989), p. 80.

Table 5.7 Direct Payments (Government Subsidies) to Agriculture, Canada and the Prairie Provinces, 1971, 1975, and 1978 to 1990 (in millions of nominal Canadian dollars)

Year	Canada			Prairies		
	Net Direct Payments[1] (1)	Net Farm Income (2)	Payments as % of Net Farm Income (1)÷(2)	Net Direct Payments[1] (3)	Net Farm Income (4)	Payments as % of Net Farm Income (3)÷(4)
1971	139	1,409	9.9	39	830	4.8
1975	479	3,977	12.1	108	2,589	4.2
1978	538	3,086	17.4	195	1,756	11.1
1979	802	3,280	24.5	364	1,648	22.1
1980	662	2,741	24.2	218	1,300	16.7
1981	857	4,005	21.4	235	2,606	9.0
1982	900	2,892	31.1	304	1,748	17.4
1983	820	2,067	39.7	219	789	27.8
1984	1,424	2,664	53.4	706	677	104.4
1985	1,879	3,381	55.6	1,202	1,601	75.1
1986	2,523	4,469	56.5	1,804	2,612	69.1
1987	3,433	4,147	82.3	2,533	1,729	146.5
1988	3,312	3,464	95.6	2,225	1,288	172.8
1989e	2,886[2]	4,925	58.6	1,921[2]	2,446	78.5
1990e	1,200[2]	2,812	42.7	632[2]	713	115.8

1 Net direct payments, which equal gross direct payments from governments to farmers minus any producer premiums or levies, do not include any indirect benefits to farmers, such as the "Crow Benefit," nor any imputation of the costs of consumer transfers to farmers.
2 Net direct payments in 1989 and 1990 were derived by using the estimates of gross direct benefits forecast by Agriculture Canada and applying the ratio of net to gross payments (88.0% for Canada, 88.7% for the Prairie Provinces) that prevailed in 1988.
Sources: Based on data in Statistics Canada, *Agriculture Economic Statistics*, Catalogue No. 21-603E, and estimates for 1989 and 1990 from Agriculture Canada, *Market Commentary* (December 1989).

Europe (assuming 8.8 million farms in the EC-12). These average figures conceal the fact that the lion's share of program benefits go to larger commercial farms in all these nations, as well as considerably underestimating the extent of support to farms in the northern half of the EC.[1] Relative to other nations, transfers in Japan are even higher.

Canada has not been immune to these pressures, as Table 5.7 attests. Direct payments from the federal and provincial governments, net of producer levies, increased markedly from 1984 onward, reaching a record high of nearly 96 percent of Canadian net farm income in 1988. The importance of direct payments to prairie farm producers in the mid to late 1980s is particularly striking, although they are estimated to have declined marginally in 1989 and may decrease substantially in 1990 in the absence of new government initiatives. By any yardstick, the extent of taxpayer and consumer transfers (direct and indirect subsidies) to farmers has grown to extreme levels in rich nations and stands greatly in need of reform.

Disruptive trade practices have also increased in the 1980s, most notably with the U.S.-EEC export subsidy war that followed the reorientation of U.S. farm policy under the 1985 Food Security Act. The heightened levels of contention and dispute relating to agricultural trade issues are also apparent in the increased numbers of consultations and dispute panels for agriculture during the 1980s under the procedures of the General Agreement on Tariffs and Trade (GATT). International tension and concern have arisen over the unilateral changes in trade policy and law, particularly with regard to features of the 1988 U.S. Omnibus Trade and Competitiveness Act. Concerns have also been expressed about the effects on international trade, both of farm products and other commodities, from the increasing number of bilateral trade agreements and the prospective integration of the EC common market by 1992. These varied pressures, together with high budgetary costs of farm policy in many high income countries, have contributed to the unprecedented focus on agriculture in the current negotiations under GATT.

An Overview of Market and Price Policies for Agriculture

Regional farm policies have been formed over many years by the interaction of fairly similar economic pressures and somewhat different social and political pressures. Thus, despite general similarities common to the agricultural policies of virtually all high-income countries, there are considerable differences in the details of the farm programs established over time in Western Europe, the United States, and Canada.

The three principles underlying the common agricultural policy (CAP) of the EC are: common prices among members (although this has been modified by the system of "green exchange rates," used in converting agricultural prices in European Currency Units to national currencies, and which have been associated with internal agricultural border taxes and subsidies); community preference; and common financial responsibility for the CAP. The institution of the common agricultural policy by the six founding members of the EC required harmonization of somewhat different systems of support and protection. In view of the long tradition of protectionism in the founding nations, especially Germany and France, it is not surprising that the CAP has, from its earliest years, involved high levels of farm prices and high levels of protection for its agricultural sector, relative both to the lower cost member states of the EC and to the rest of the world. Consequently, the major source of producer-income support is through price-based transfers from consumers; these have accounted for more than 80 percent of transfers to producers in recent years (Table 5.6).

Originally, the EC was not self-sufficient in many farm products, but this has changed as high internal price guarantees, applied until recently without production limits, have encouraged production from higher-cost regions and farmers, leading to considerable costly surpluses and intervention stocks. In general, community preference has been expressed, and guaranteed price levels maintained, by variable import levies coupled with export refunds. This combination gives a system that cushions the EC from variations in external prices arising from changes in the rest of the world. It also enables variations in internal supply and demand that might otherwise lead to internal price instability being transferred to the rest of the world. As domestic production levels have increased, the contribution from variable import levies to the EC budget has diminished. Meanwhile, the budgetary costs associated with EC agricultural market and price policy have greatly increased, particularly for the various intervention mechanisms and export refunds. Budgetary pressures intensified in the 1980s due to depressed world prices and payments for the new members, Spain and Portugal. Efforts to cap these escalating budgetary costs underlie changes in EC farm programs that have occurred since the mid-1980s. The most significant change was the introduction of dairy quotas in 1984. Subsequently, since 1987, "stabilizer" mechanisms have been introduced to reduce effective levels of farm prices when aggregate output increases beyond specified levels ("maximum guaranteed quantities").

In contrast to farm policy in the EC, producers in both the United States and Canada, with some exceptions, receive relatively more support through transfers from taxpayers than from consumers. The exceptions, for the

U.S., are sugar and dairy products; for Canada, dairy and poultry products. For these commodities, restrictions on imports have enabled domestic prices to be maintained at relatively high levels, although rather different approaches have been followed in these and other programs for agriculture in the two countries.

Levels of price and income support for U.S. agriculture have varied over time, changing in response to the economic and political environment. They have been considerably higher in the 1980s than in the preceding decade. The major features of current U.S. farm programs are expected to be retained in the next farm bill. As in the EC and Canada, the support for different commodities varies considerably, reflecting long-standing differences in political and economic characteristics. Since 1985, assistance for the U.S. crops sector has focused on deficiency payments that bridge target prices and either market prices or loan rates, the mechanism of providing floor prices. Deficiency payments have been tied to acreage reduction provisions, which are intended to reduce the stimulating effects of the programs on agricultural production.

Export subsidies for farm products were re-introduced under the 1985 farm bill. These "payment in kind" bonuses on particular sales to specified markets relate to disposal of government financed stocks. The original targeting of export enhancement subsidies to regions where the U.S. felt it had suffered unfair competition from EC export subsidies soon became much broader, and is believed to have resulted in considerable downward pressure on grain prices in world markets. The program has also been applied to other products, including eggs and poultry. Import quotas for sugar and dairy products, and the use of voluntary export restraint agreements to reduce U.S. meat imports from Australia and New Zealand, have been the major means of providing support, through higher consumer prices, for these products.

Canada has a variety of programs and institutions relating to markets and prices for agriculture. This is due partly to the sharing of responsibility and regulatory power over these areas by the federal and provincial governments, and partly to regional diversity within this country. Programs have evolved incrementally; they have often been introduced in times of crisis and persist long afterwards. As a relatively small open economy, with a western farm sector located far from major markets, transportation policy and subsidies have been emphasized for grains and oilseeds. Supply management systems involving relatively high prices and quota restraints on entry, production, and imports have been developed for the dairy and poultry sectors, which tend to be more concentrated near the major domestic markets in Ontario and Quebec. Levels of support for milk producers, largely from consumer transfers but also from direct subsidies, are higher than in either the US or EC. Stabilization programs, funded partly by participating farmers, are intended to provide a safety net against declining prices for grains, livestock, and some other products; payments from these have been substantial for some commodities in recent years, especially for grains. These stabilization payments and other programs have been treated as countervailable subsidies in recent U.S. trade law decisions.[2]

Historically, the extent of protection and support accorded the Canadian agricultural sector was relatively modest compared to that provided in many other developed countries. Over the past twenty years, however, the trend toward protection and support has increased, with supply management programs giving considerable support to dairy and poultry farmers, and provincial governments becoming increasingly involved in a variety of farm stabilization and support programs, and, most recently, offering substantial support to grains and oilseed producers in reaction to low world prices and drought. This trend is reflected in Table 5.6. Canada now has a relatively highly supported and protected agricultural sector. The degree of support to western grain and oilseed producers is expected to fall in 1990, and there are considerable political pressures for further transfers.

Pressures for Reform in Western Europe and North America

There are pressures for reform of the commodity-oriented market and price support systems for agriculture in Western Europe, the United States, and Canada, but the strength of these and the underlying economic and political interests differ by region and commodity. Budgetary pressures arising from the high costs of farm programs are evident in each. Demands to reduce these costs may be strongest in Canada, in view of the relatively small population base and current concerns regarding the levels of government expenditure and the national debt. Nevertheless, there is still considerable pressure to maintain government support for the western Canadian grain sector in view of the persistence of low world prices, and the Canadian government has been extremely reluctant to move against the interests of the supply-managed dairy and poultry sectors. Overall, however, the success or failure of the Uruguay Round agricultural trade negotiations will depend essentially on the willingness of the United States and European Community to compromise on agricultural trade and related farm policy issues.

The most significant pressures for reform of EC farm and agricultural trade policy arise from internal budgetary issues relating to the common agricultural policy, and pressures outside agriculture demanding agreement in other areas in the current round of multilateral negotiations under GATT. Concerns relating to budgetary costs do not focus on a decrease in current levels of support for farm prices and markets, but on efforts to restrain their rate of increase. There are concerns about the impact on the budget of supporting the agricultural sectors of new EC members, and the political problems of conflicts arising because of the varying levels of CAP costs and benefits in different member states. The EC position in policy reform and agricultural trade negotiations results from pressures of its own member states, with the British and Dutch demanding somewhat lower levels of support and German interests most opposed to this. Since the CAP has been traditionally viewed as a cornerstone of the European Economic Community, and its maintenance is considered to be basic, the EC is extremely defensive about international trade reform, and would have preferred movement toward managed trade arrangements for "problem" commodities as a solution to agricultural trade difficulties. Nonetheless, as well as suggesting international commodity agreements, especially for stocks, as a means of moving toward more "balanced" agricultural markets, the EC proposed negotiations on significant reductions in agricultural support, including subsidies as well as frontier measures. By December 1989, although reluctantly, they also agreed to discuss a form of tariffication of variable import levies involving fixed and variable components, subject to consideration of rebalancing (EC, 1989). This proposal represented substantial modification of their earlier insistence on negotiating short-term measures to ameliorate major commodity problems and conflicts.

The EC has proposed that negotiations to reduce support be in terms of an aggregate measure of support, a "support measurement unit" that would capture the trade-distorting effects of domestic agricultural subsidies and deficiency payments as well as border measures. The proposal would give credit for the EC's reductions in output achieved by quotas for milk and sugar, and relate to fixed external prices in order to avoid the influence of fluctuating world market prices and exchange rates. The use of such a measure to negotiate reductions in support is also consistent with the EC desire to "rebalance" agricultural support. Political pressures have been building within the EC to limit imports of cereal substitutes (used in animal feeds) as well as oilseeds and oilseed products. Import duties for these were bound at low or zero levels in earlier GATT rounds. Thus, current GATT negotiations provide an opportunity for rebalancing agricultural support, by increasing protection levels for these products in exchange for lower levels of protection on other farm commodities.

Canada's pursuit of global reform of agricultural policy and trade largely reflects its position as a small nation with major agricultural export interests, primarily for grains and oilseeds, but also for some livestock commodities. With the other members of the Cairns group, Canada has been concerned that agricultural trade be brought completely within GATT by eliminating exemptions from GATT procedures and measures not provided for in that agreement (Cairns, 1989). This would eliminate waivers (such as the U.S. Section 22 waiver)[3] and bring the wide variety of existing measures not covered by GATT, including nontariff barriers, variable import levies, and agricultural export subsidies, into the agreement, under new GATT rules. The Cairns group proposes substantial reductions in import-access barriers paralleled by reductions in export subsidies and internal support measures. Like the U.S. (1989), the Cairns group has pursued proposals to classify subsidies into three categories. Some would be prohibited; some would be allowable but subject to disciplines; and others would be allowable and noncountervailable. In common with the U.S. position, export subsidies would be phased out and disallowed. Trade-distorting support, whether by price support or deficiency payments, would be reduced by annual commodity-specific cuts in producer support prices and budgetary expenditures. Nontariff trade restrictions would be converted into tariff equivalents, which would be negotiated downward, and bound. The question of import restrictions under Article XI:2(c), invoked by Canada in support of its supply management programs, has been a point of contention within the group. Except for Canada, Cairns group members, like the US, would phase out these import restrictions. The wording on import access in the Cairns comprehensive proposal, tabled in November 1989, refers to "all measures not explicitly provided for in the GATT," and thus side-steps these Article XI import restrictions, a considerable concession to Canada's position on this issue. In response to effective lobbying pressures exerted on the Canadian government by Canadian supply-management commodity groups, particularly those acting for the dairy industry, Canada subsequently tabled a proposal to clarify and extend provisions of Article XI, a position that has support in other nations where agriculture is highly protected; these include Japan, South Korea, the EC, and the Nordic group of countries.

The pressures in the U.S. for global reform of agricultural trade policy reflect that nation's position as a large economy with major trading interests, and an ideological leaning toward open markets. Nonetheless, there are powerful political pressures for protectionism in the U.S., including some sectors of agriculture. As in other large nations, budgetary constraints place some restrictions on farm programs, but these can be overweighed by political pressures. The U.S. wants to reduce market-access barriers for grains, oilseeds, citrus, and some livestock products. It has pursued bilateral negotiations and trade arrangements to this effect. It has also pursued an aggressive policy of export subsidization, a reaction to the fall in export market shares and increased grain stocks from 1981 to 1985, the inevitable outcome of the relatively high levels of loan rates under the 1981 farm bill. As indicated by OECD (1989) calculations of the producer subsidy equivalent measure, protectionist pressures are most evident for sugar and dairy products, and are also appreciable for beef. Compromise in the agricultural trade negotiations may be aided by U.S. interest in the other major areas of the Uruguay Round multilateral trade negotiations (trade in services, trade related to investment measures, and intellectual property).

Overall, trade liberalization and the associated reform of farm policy is expected to yield real income gains. But there are political costs connected with anticipated changes in income that would flow from associated changes in farm programs. The benefits would be widely spread, and the costs would be more narrowly concentrated, a feature that has, until now, discouraged reform. Overall, the major beneficiaries of policy reform would be consumers, taxpayers, and, depending on current levels of support, low-cost producers. Higher cost and highly protected or supported producers expect income losses as a result of lower support prices and/or government payments, with consequent losses in asset values. Such groups are well-organized and effective lobbyists. Nonetheless, international and national economic welfare could be enhanced, with compensation to those disadvantaged by change, by improved policies for agriculture in rich nations. The forces for agricultural trade policy reform were insufficient for meaningful change in previous GATT rounds. In this round, agreement on reform may occur to some degree in spite of, rather than because of, pressures from within agriculture, since the pressures for reform are linked to wider issues in the GATT negotiations. In addition, the political will to pursue farm policy reform is likely to be greater, and the political costs less, in the context of multilateral rather than unilateral reform. This may explain the unprecedented levels of national political commitment to pursue reform of agricultural trade. Currently, and in the near future, changes in national policies for agriculture are likely to be greatly affected by the nature and extent of progress in the multilateral trade negotiations.

Policy Options for the Future

The crisis facing European and North American agriculture is not insurmountable. Many economists, over many years, have outlined the general policy changes needed if reform is to occur (see, for example, "Statement of Twenty-nine Professionals from Seventeen Countries," 1988). What has been lacking, for the most part, is sufficient political will, at both the national and international levels, to implement the necessary changes in domestic agricultural policies and in international agricultural trade policy.

Any discussion of policy options must be couched in terms of a vision for the future of agriculture, particularly the current and emergent economic forces that impinge on the agricultural sector. Real commodity prices in agriculture are expected to continue on a gradual downward trend, some of which, as for wheat and corn, have been evident for several decades. A secular decline in real prices is the outcome of Engel's Law (as incomes rise, a smaller proportion is spent on food) and continuing productivity growth in agriculture (in recent decades, annual increases in total factor productivity have ranged from 1 to 2 percent in the United States and Canada). A Malthusian era of chronic world food scarcity is extremely unlikely. World food production is expected to continue to grow more rapidly than the world's population, currently increasing at 1.8 percent per year. The developing nations, however, will likely continue to import more food into the next century, as their annual rate of growth of effective demand (currently approximately 3 percent) continues to outpace the rate of growth in food production, which has been roughly 2.5 percent annually over the past three decades.

These various factors offer numerous implications for the future of agriculture in developed nations. There are likely to be fewer farms, increasing farm size, gradually declining numbers of farmers, greater integration of agriculture with the nonfarm economy (with off-farm income being an even more significant determinant of farm families' income), and greater exposure of the farm sector to both national and international macroeconomic forces. These are not new forces, and are likely to continue into the future. In short, there will be continuing pressure for resources, especially labor and land, gradually to leave agriculture in rich nations. A basic choice, then, is whether agricultural policy in developed nations should be designed to facilitate or resist such fundamental change. There are strong economic arguments for a set of policies for agriculture in Europe, North America, and Japan that rely more heavily on market forces to allocate the resources of labor and land, but which also target economic assistance to needy farm families, and provide government programs for areas where markets do not work well. This would mean relying on agricultural price and trade policies that are more market-oriented, more outward looking, and less protectionist than currently in existence. It would also require targeting farm support, rather than applying it indiscriminately through measures related to prices of outputs, costs of inputs, or interventions in domestic and international trade. It would include measures to assist farmers to manage risks, and more intervention in agriculture where markets do not function well—in some areas of environment or food safety, for example.

The spectrum of policy choices in agriculture in developed nations ranges from complete reliance on free markets (with little concern for distributional, environmental, or risk-management issues) to mandatory supply control (see Table 5.1). The latter has been seen as in the producer interest (particularly for producers in high cost sectors), although this ignores the effects of increased costs related to the inflated asset values that have accompanied effective supply control. Thus, the programs have been in the interests of "first generation" producers, those in the industry when the program was introduced, but not in the wider public interest. These types of programs have sometimes been argued for on the basis that agriculture faces oligopoly in most input markets and some output markets. This rationale ignores the fact that while countervailing power may be effective where two opposing forces are involved, given a third linked but unconcentrated sector (consumers), producers and processors tend to jointly exert, against that group, the market power bestowed by supply control.Further, the existence of concentrated input and output markets for processors and input suppliers does not necessarily imply a lack of workable competition. The competitive pressures stemming from imports and the pursuit of an effective competition policy are likely to be more

Figure 5.1 The Range of Policy Choices in European and North American Agriculture

Less Intervention				More Intervention
<				>
Free markets	Free markets with risk management	Free markets with decoupled payments	Current programs	Mandatory supply controls

Source: U.S. Department of Agriculture, *Agricultural-Food Policy Review: U.S. Agricultural Policies in a Changing World* (November 1989), 290.

beneficial to the public interest than protecting agriculture, because associated industries are more concentrated than agriculture.

It has also been suggested that North American agriculture could unilaterally or bilaterally exert market power in agricultural exports, via, for instance, a wheat export cartel. This suggestion has met with scepticism since it is highly likely that high-cost production would be encouraged elsewhere, undermining the possible benefits of cartelization (Antle, 1988). It is possible that one outcome of current negotiations within GATT could be for a system more akin to "managed trade," than multilateral free trade; this would be a second-best option, possibly reducing instability, but without the efficiency and income benefits of freer trade.

Commercial agriculture in Europe, the United States, and Canada must be increasingly subjected to the pressures of the market—that is, to the price signals arising out of the longer-run trends in supply and demand for agricultural products on a global basis. In general, agriculture in rich nations has had many sectors with prices supported above market-clearing levels, while agriculture in poor nations has too often had agricultural prices depressed below equilibrium levels. By using price policy indiscriminately for distributional purposes, governments in rich nations have unduly distorted output and resource allocation decisions, leading to economy-wide inefficiencies and barriers to international trade.

There are, of course, costs involved in moving from current farm programs to a more market-oriented system. These include: (1) the short-run losses in farm income as support is withdrawn from subsidized sectors; (2) the loss in the value of farm assets and quota rights (where the benefits of farm programs have been capitalized into such assets); and (3) the costs of adjustment to alternative employment (Johnson et al., 1985). Temporary government assistance, including lump-sum transfers to compensate in part for losses in asset values, may be warranted to cushion these costs of adjustment and to make the movement to a more market-oriented system politically feasible.

While market or price systems are useful, decentralized, administrative arrangements for allocating resources in both capitalist and socialist nations, they are not without limitations. Government intervention is required to rectify inequity in the distribution of income, to deal with areas of market failure such as externalities (including environmental problems) and the provision of public goods (including some types of research and education), and to cope with risk and uncertainty in the absence of complete markets for the handling of risk.

Regarding income distribution, there is widespread concern that farm programs in Europe, the United States, and Canada have not reduced income disparity within agriculture, and their benefits have not reached the farm families who are most disadvantaged. Farm support has typically been across the board, on a per-bushel or per-acre basis, to all farmers, insufficiently discriminating the "rich" from the "not-so-rich" or "poor" in the sector. Programs that deliver benefits on the basis of inputs or outputs will widen, rather than narrow, disparities in farm income. In most rich nations, three distinct types of farmers are emerging: part-time farmers, who rely increasingly on off-farm income to generate a living; low-resource, low-income farmers, who rely primarily on agriculture for their source of income; and commercial farmers, who produce the bulk of national agricultural output. Income aid must be concentrated on those farm families most in need (primarily in the middle group), through decoupled payments, whether by targeted programs or, perhaps, guaranteed income schemes. Public policy can also assist in facilitating the transfer of low-income farm families to nonfarm employment.

The market system also fails to deal with environmental problems. Citizens in both Europe and North America are becoming increasingly concerned with environmental issues, including those relating to agriculture. The current watch-word of "sustainable development" carries the implication that farming practices be sustainable over time, and not degrade the resource base irreparably nor impose external costs on other sectors. In Europe, with higher population densities and intensive farming systems, there are ongoing concerns with the provision of green-belt amenities and the handling of agricultural pollution associated with intensive farming. Excessive price levels appear to be encouraging high levels of chemical application and overintensive farming methods. In the United States, the foremost environmental problem in agriculture currently relates to ground water quality and contamination. Environmental problems in Canada range from water quality in the Great Lakes to soil quality degradation of the prairie land base.

In all these nations, environmental dimensions of agricultural policy will be much higher on the political agenda in the 1990s. In fact, there may be scope to integrate environmental objectives with domestic policy reform (OECD, 1989a). In the "swamp-buster" and "sod-buster" provisions of the 1985 Farm Bill in the United States, conservation objectives were married to the continuance of commodity support. In both Europe and North America, there is scope for the development of imaginative programs that link payments to farmers for the provision of extra-market benefits (such as habitat retention or green-belt amenities) to elements of domestic agricultural policy reform. In Europe, for instance, there are some, albeit minor, environmental provisions in the movement to further integration in "Europe

1992," which offer a rationale and a starting point for providing European farmers with decoupled income payments (USDA, 1989). In addition to environmental problems, the food safety concerns of consumers will rise in political importance, and governments will assume an enhanced role in research and education relating to these concerns, as well as in regulating and monitoring programs that deal with them.

Government, of course, must play a considerable role in the provision of goods and services, such as research and education, two high pay-off areas of societal investment. Improved basic education and enhanced training in rural areas are critical to both the farm families who remain on the land and those who move to nonfarm employment.

Finally, there is an apparent need for governments to assist farmers in managing risk, as markets to share, trade, and reduce risk are incomplete. This typically has taken the form of crop (yield) insurance, price stabilization, and income stabilization programs. Price stabilization programs often involve price support, but if their focus is truly that of providing a safety net, their level should be below, rather than above, the expected market-clearing level. Income stabilization plans could be primarily or fully based on producer contributions, as is currently the case in New Zealand and Australia, so that they do not distort production or trade. Only a small proportion of North American farmers use forward markets, so there is scope for extension and training in this area, by private firms, marketing co-operatives, and government.

In conclusion, there are convincing economic arguments to support the case that output levels, input use, price levels, and trade flows in NorthAmerican and European agriculture should be increasingly subjected to market-related pressures. There are crucial roles for government in the agriculture of the future: ensuring equity in income distribution, managing environmental problems, helping provide research, education, and training, and assisting in reducing price and income risk.

Conclusion

Current political and economic circumstances combine to provide an unusual opportunity for international and national reform of farm and trade policy. If reform occurs in the context of the multilateral trade negotiations, it is likely to provide for gradual, rather than abrupt, change. And there are real risks that these changes may only be marginal. The view of agriculture as deserving special attention is well entrenched in North America and

Figure 5.2 Some Prevalent Myths that Relate to Agriculture in North America and Western Europe

---

Myths Common to All These Regions:

1. Farming is a major contributor to national income and employment.
2. Support and protection of agriculture is necessary to ensure the nation does not run out of food, and to feed the world's starving millions.
3. Farmers' incomes are falling further behind those in nonfarm occupations.
4. Farmers are politically disadvantaged; their dwindling numbers leave them with little political power.

Some Country-Specific Myths:

5. U.S.: All other nations but the U.S. subsidize and support agriculture and farm exports.
6. Canada: Canada has an open, nonsubsidized agricultural sector.
7. EC: EC agricultural protection and support has fallen to very low levels.

The Realities Are:

1. Farming is a major component of some regional economies, but represents a relatively small and declining proportion of total output and employment of the national economy.
2. The United States and Canada are important net exporters of food. The EC is self-sufficient in many farm products. Policies to support farmers in high income countries have depressed food prices globally and discouraged agriculture in developing countries.
3. Farm family income has increased significantly, relative to income levels of all families, over the past two decades in North America.
4. Rural electoral districts include fewer voters than urban areas, giving farmers proportionally stronger political power.
5. U.S. levels of protection and support for agriculture have been substantial in the 1980s.
6. Average producer subsidy equivalent measures for agriculture in Canada in the 1980s exceed those in the U.S. and approach those in the EC.
7. Considerable protection and support is afforded EC farmers.

Western Europe, as shown by the pervasive nature of the beliefs noted in Figure 5.2.

If meaningful reform is not achieved in the context of the multilateral trade negotiations, should reform of farm policy be pursued unilaterally? Such a course of action would provide economic benefits for the national economy, as would the unilateral reform of trade policy. Hertel (1989) concludes that unilateral trade policy liberalization of US agriculture would lead to substantial benefits across the economy. The particular economic and political factors underlying the decisions of Australia and New Zealand to unilaterally disengage themselves from the international war of farm subsidies and protection from trade pressures, deserve further study. The political and economic costs of farm and trade policy reform are expected to be much less, however, if reform occurs in a multilateral framework.

The crisis in European and North American agriculture can be solved by means of more rational farm policies. Such policies would be more outward-looking, less oriented to supporting prices above market-clearing levels, and focused on areas where market forces do not work well—primarily income distribution, risk, and environmental issues.

Notes

1. On a per-hectare basis, the extent of transfers in 1988 averaged approximately US $900 per hectare of cropland and pasture in Europe, a level more than double that in the United States and more than five times that in Canada (using only cropland and not permanent pastures in the calculations for the United States and Canada).
2. The U.S.-Canada Free Trade Agreement, signed in January 1988, has not had major effects on agricultural trade or on the farm policies in either nation; the most contentious issues were deferred to multilateral negotiations within GATT. However, under the U.S.-Canada bilateral agreement, tariffs on farm and other goods are being reduced over a ten-year period. Supply managed commodities are largely bypassed by the agreement. The agreement provides for monitoring support levels for grains and the elimination of Canadian import licensing for these products, as U.S. levels of support for the grains sector fall to the Canadian level. It is anticipated that the dispute settlement mechanism of the agreement will streamline the application of each nation's countervail and antidumping laws.
3. This is a 1955 GATT waiver of U.S. obligations, enabling that nation to restrict imports in support of price support operations under Section 22 of the U.S. Agricultural Adjustment Act.

References

Anderson, Kym, and Yujiro Hayami. *The Political Economy of Agricultural Protection.* Sydney: Allen and Unwin, 1986.
Antle, John M. *World Agricultural Development and the Future of U.S. Agriculture.* Washington, D.C.: American Enterprise Institute for Public Policy Research, 1988.
Cairns Group. "Comprehensive Proposal for Long-Term Reform of Agricultural Trade." November 1989.

Commission of the European Communities. *The Agricultural Situation in the Community: 1988 Report.* Luxembourg: Office for Official Publications of the EC, 1989.

Economic Council of Canada. *Handling the Risks: A Report on the Prairie Grain Economy.* Ottawa: Minister of Supply and Services, 1988.

European Community. "Agricultural Comprehensive Proposal for the Long Term in the Punta del Este GATT Round." December 1989.

Fulton, Murray, Ken Rosaasen, and Andrew Schmitz. *Canadian Agricultural Policy and Prairie Agriculture.* Ottawa: Economic Council of Canada, 1989.

Gilson, J.C. *World Agricultural Changes: Implications for Canada.* Toronto: C.D. Howe Institute, May 1989.

Hathaway, Dale E. *Agriculture and the GATT: Rewriting the Rules.* Washington, D.C.: Institute for International Economics, 1987.

Hertel, Thomas W. "Domestic Implications of Trade Policy Liberalization." In U.S. Department of Agriculture, *Agricultural-Food Policy Review.* Washington, D.C.: USDA, 1989: 275-88.

Johnson, D. Gale, Kenzo Hemmi, and Pierre Lardinois. *Agricultural Policy and Trade: Adjusting Domestic Programs in an International Framework.* New York: New York University Press, 1985.

Kramer, Carol S., ed. *The Political Economy of U.S. Agriculture.* National Centre for Food and Agricultural Policy: Annual Policy Review 1989. Washington, D.C.: Resources for the Future, 1989.

Organization for Economic Co-operation and Development (OECD). *Agricultural Policies, Markets and Prices, Monitoring and Outlook, 1988.* Paris: OECD, 1988.

———. *Agricultural and Environmental Policies: Opportunities for Integration.* Paris: OECD, 1989a.

———. *Agricultural Policies, Markets and Prices, Monitoring and Outlook, 1989.* Paris: OECD, 1989b.

Policy Statement by Twenty-Nine Professionals from Seventeen Countries. *Reforming World Agricultural Trade.* Washington, D.C.: Institute for International Economics, and Canada: The Institute for Research on Public Policy, May 1988.

Schmitz, Andrew, Alex F. McCalla, Donald O. Mitchell, and Colin Carter. *Grain Export Cartels.* Cambridge, Mass.: Ballinger, 1981.

Statistics Canada. *Agriculture Economic Statistics.* Cat. 21-603. Ottawa: Minister of Supply and Services, semi-annual updates.

Tracy, Michael. *Government and Agriculture in Western Europe 1880-1988.* 3rd ed. New York: New York University Press, 1989.

U.S. "Submission of the United States on Comprehensive Long-Term Agricultural Reform." October 1989.

U.S. Department of Agriculture. *Agricultural Statistics 1988.* Washington, D.C.: U.S. Government Printing Office, 1988.

U.S. Department of Agriculture, Economic Research Service. *Agricultural-Food Policy Review: U.S. Agricultural Policies in a Changing World.* Agricultural Economic Report No. 620. Washington, D.C.: USDA, November 1989.

———. *Western Europe: Agriculture and Trade Report.* RS-89-2. Washington, D.C.: USDA, July 1989.

World Bank. *World Development Report 1989.* New York: Oxford University Press, 1989.

6

The GATT: Its Historical Role and Importance to Agricultural Policy and Trade

Tim Josling

Introduction

The multilateral trade system is facing its most significant challenge in recent years. The attractions of bilateral or regional trade pacts have been given a boost by the European Community's (EC) imaginative initiative to "complete" its internal market by the end of 1992. There is renewed talk of Pacific Rim trade associations. The U.S./Canada Free Trade Agreement, though not without its problems, has shown that like-minded countries can put together a package of trade liberalization in a relatively short period of time. Unilateral commercial diplomacy has also proved seductive, particularly in the U.S. Congress, where the temptation for one country to set rules and act as judge and jury is being encouraged by the "Super 301" provisions of the 1988 Omnibus Trade Act.

All this is taking place against the backdrop of a remarkably wide-ranging set of negotiations in the General Agreement on Tariffs and Trade (GATT)– the Uruguay Round. If successful, the GATT will have extended its influence to include trade in services and international aspects of property rights and investment policies. It will have strengthened its rules on nondiscrimination in government purchasing contracts, overhauled its dispute-settlement mechanism, and forged ties with the multilateral monetary and financial institutions. Among its most ambitious tasks, however, is to revise the rules of agricultural trade, making them apply with greater force and clarity to domestic agricultural policies. If this can be achieved, the Uruguay Round will rewrite the book on agricultural trade policy and influence farm product markets for years to come. But success in the Uruguay Round is by no means assured, and an outright failure could lead more people to agree with Lester Thurow that "GATT is dead," or with Susan Strange, who has stated that "if GATT were to slip below the water of Lac Leman, no one would notice."

The implications of success and failure are the subject of Professor Runge's chapter, "Prospects for the Uruguay Round in Agriculture." The task of this chapter is to provide some background to the GATT, and to the Uruguay Round negotiations in agriculture. An attempt will be made to stick to the historical perspective, with the understanding that "history" includes the comprehensive proposals of October to December 1989.

The GATT

The General Agreement on Tariffs and Trade has been described as an orphan among international institutions. It was negotiated in 1947 as a commercial code of conduct, part of a broader International Trade Organization (ITO). The treaty setting up the ITO (the Havana Charter) was never ratified by the U.S. Congress, who considered the charter's sweeping authority over trade policy too restricting on its own actions. The GATT slipped through on a technicality, as the U.S. president has authority to negotiate and conclude "executive agreements" in the trade area.[1] Though there have been recent suggestions that the ITO be revived, GATT has held up remarkably well as a low-key, small-secretariat institution with a limited and provisional mandate in international affairs.

The GATT provides a set of rules which the "contracting parties" agree to observe; acts as a forum for the negotiation of changes in trade barriers; and provides a means of settlement of disputes. The basic propositions embodied in the GATT are plainly liberal in intention, emphasizing the mutual benefits of freer trade, outlawing export subsidies and quota restrictions in most cases, and encouraging mutual reductions in bound tariffs. The GATT preaches nondiscrimination among member states, granting market access on the basis of "most-favored nation" status to all other contracting parties—subject to defined exceptions for customs unions and developing countries. It allows for the settlement of disputes, for compensation for the impairment of benefits, and ultimately for trade sanctions against violators of the agreement. Though originally oriented largely toward western industrial countries, its membership has now expanded to ninety-six nations, and includes many developing countries and some central European economies. Seven rounds of negotiations have been successfully completed since the inception of the GATT, which have led effectively to a low-tariff system for manufactured products in developed-country markets. The eighth such round, the Uruguay Round, has been underway since September 1986.

Agriculture in the GATT

From the beginning, agriculture proved a problem area for the GATT system of more liberal trade. Quantitative import restrictions on agricultural goods were excluded from the agreement's strictures against nontariff restrictions on trade (Article XI), provided that the country in question practised supply control. Variable levies are not presently mentioned in the GATT, and hence fall outside the GATT disciplines.[2] Primary products were explicitly excluded from the general prohibition on export subsidies (Article XVI). Although other parts of the agreement do require that governments manage their domestic agricultural policies so as not to harm the legitimate interests of others, and oblige governments to consult on agricultural trade problems, these provisions have been rarely applied. Waivers were frequently sought to exempt particular agricultural policy actions from the GATT code. Most notable among these derogations was the waiver granted to the United States in 1955, which effectively removed the major elements of U.S. agricultural policy from international scrutiny.[3] This insistence by the U.S., as the guiding force in the early years of GATT, that domestic policies should not be subject to international limitations, has severely hampered attempts to liberalize trade in farm products. More recently, the tacit acceptance by most countries of the Common Agricultural Policy (CAP) of the European Community, even though its major instruments have uncertain legality in the GATT, has reinforced the notion of the primacy of domestic policy in international trade discussions.

One measure of the unsatisfactory nature of agricultural trade rules is the number of complaints brought in the GATT in this area. Since 1976, there have been nineteen occasions when a GATT panel has been appointed to look into an agricultural trade complaint, out of a total of thirty-two GATT cases over this period. Twelve of these nineteen complaints involved objections against the EC, and included products ranging from pasta and wheat flour to poultry, citrus, beef, apples, sugar, and animal feed. Two reasons stand out as to why the EC is so often cited by trading partners: the widespread use of targeted export subsidies (refunds), relatively unconstrained by spending curbs; and a switch from deficit to surplus in many products, unaccompanied by a sufficient reduction in support prices to take account of that change. As a consequence, EC export (and import) policies have been under attack in the GATT for two decades. But the impact of these actions on the CAP has been minimal. Indeed, until recently, it has been difficult to find any examples of domestic policy changes stemming directly from GATT rulings.[4]

Previous Negotiation Rounds

The ineffectiveness of the GATT regulations in dealing with illiberal elements of national agricultural policy has not prevented the incorporation of agricultural discussions in the recent GATT rounds of trade negotiations. Though domestic policies have (until now) been kept off the agenda, a modest degree of liberalization has been achieved through the reduction of tariffs and the establishment of tariff-free quotas for certain products. A number of agricultural tariffs were "bound" in the Dillon Round of trade talks (1960-1961), when the newly formed European Community negotiated the necessary changes in member-state duties on imports from third countries. Some of these commodities, such as soybeans, high-protein meals, and cassava, later became big-ticket items in international trade—much to the discomfort of EC policy-makers and to the credit of the GATT.[5]

The isolation of agriculture from the GATT began to change in the mid-1960s. The U.S. became convinced that it had a comparative advantage in agricultural products, and that this profitable trade was being hampered by protection in other countries. In the Kennedy Round (1964-1967), a major discussion about the nature of agricultural trade policy took place, with the United States arguing for a return to the GATT notion of a market-oriented trading system, and the European Community generally favoring managed markets through commodity agreements. This philosophical difference was never solved, and echos of it are still evident today.

At one stage in the negotiations, the EC proposed a temporary binding of support levels, relative to negotiated reference prices (the "montant de soutien" proposal), but this was rejected by the United States as perpetuating rather than removing the protectionist policies in importing countries. Finally, an "agricultural component" of the Kennedy Round was put together, which included in particular an International Grains Arrangement (IGA) (1967), a modified version of earlier International Wheat Agreements. The market stabilization provisions of the IGA had a short life, as surplus grain production soon pushed prices below the established minima, and grain-producing countries proved unwilling to hold either exports or domestic output in check.[6]

In the next round of talks, usually known as the Tokyo Round, which concluded in 1979, agriculture again appeared high on the agenda. The underlying issue was still the extent to which agricultural trade should be subject to the same rules as trade in manufactured goods. Should markets be free or managed? Should domestic policies be the subject for negotiations? Should export subsidies be banned or otherwise controlled? Should quantitative trade barriers be phased out? And, should importers be able to shield their markets from disruption? Disagreements on these fundamental issues persisted through the Tokyo Round, and prevented any substantial progress toward liberalization. Some quantitative restrictions were relaxed (in particular by Japan), and two commodity agreements were concluded (for dairy products and for meat), neither endowed with the instruments to stabilize markets. At the same time, talks proceeded on another International Wheat Agreement, once more serving as an agricultural "component" in the round, though in reality the product of a separate set of negotiations. It suffered a worse fate than the IGA: it proved unpopular with developing countries and never saw the light of day.

One development of the Tokyo Round, the attempt to refine the rules governing export subsidies and countervailing duties, held out some promise for agricultural trade. The Subsidies-Countervail Code obliged countries to avoid export subsidies that lead either to an "inequitable" share in world markets or undercut prices. However, the implementation of that code, through the challenge within the GATT of certain practices (such as the EC's export refunds), has not been successful. Interpretation of such concepts as an "equitable" market share lead inevitably to conflicts and can render meaningless the spirit of the original agreement.[7]

Whether the GATT can improve its effectiveness in dealing with agricultural issues is being put to the test in the current GATT round of negotiations, the Uruguay Round, which includes agriculture as a centerpiece. How far this initiative can proceed depends upon whether the countries concerned, particularly the EC and the U.S., have confidence in a multilateral mechanism for resolving trade issues in agriculture.

Lead-up to the Uruguay Round

International agencies have little influence over domestic policies under normal conditions, but every now and then political and economic circumstances conspire to produce a demand for the product of the international bureaucracy. The period from 1982 to 1986 was such a time. The EC and the U.S. watched as their agricultural spending rose almost out of control; the U.S. lost much of its market share painfully accumulated over the 1970s; and even Canada began to reverse a trend toward more liberal policies. Both the GATT and the Organization for Economic Cooperation and Development (OECD) were at hand to offer solutions to the problems. Their initial responses were characteristically different. In November 1982, the GATT Council established a Committee of Trade in Agriculture (CTA) to look into the ways in which agricultural trade could be better handled within the GATT framework. Many of their proposals, tabled in November 1984, addressed the issue of the way in which existing GATT articles and codes could be applied more effectively. Particular attention was given to the implementation of Article XI, which allows quantitative restrictions on agricultural imports under certain circumstances, and Article XVI, which condones export subsidies for primary products.

Meanwhile, a parallel activity was going on in the OECD. Agricultural policies in OECD member countries have been studied by the secretariat in the past and have been discussed in the Agriculture Committee. But unlike macroeconomic policy, where such discussions of national policies have led to a degree of coordination, OECD discussions of agricultural policies have rarely gone beyond the descriptive. The OECD ministerial meeting in May 1982 gave the secretariat a mandate to undertake "an analysis of the approaches and methods for a balanced and gradual reduction in the protection of agriculture ... an examination of relevant national policies and measures which have a significant impact on agricultural trade ... [and] an analysis of the most appropriate methods for improving the functioning of the world agricultural market."[8] This initiative, known as the Trade Mandate, was to be supervised jointly by the Committee on Agriculture and the Trade Committee. This joint committee reported the results of its study to the Ministerial Council of the OECD in May 1987.

The OECD interpreted the mandate in an appropriately broad way, and looked at the whole range of domestic agricultural policies. Building on a quantitative measure of farm support used previously in the Food and Agricultural Organization (FAO), the OECD calculated Producer Subsidy Equivalents (and Consumer Subsidy Equivalents) for several member countries. Though countries with high levels of protection were less than enthusiastic with this attempt to quantify the effect of support policies, the Trade Mandate report represented a milestone in the discussion of agricultural policies and trade. In what constituted an important political development, details of national policies were openly discussed for the first time in a quantitative framework, and trade impacts were examined in an even-handed way. Subsequent decisions to update the calculations have suggested that countries find them useful as information for improved policy decisions. 9

The OECD ministers, in May 1987, confirmed their commitment to trade negotiations, but went further in advocating changes in domestic programs toward a greater market orientation. The ministers called for concerted efforts to reform domestic farm policies, with the long-run aim of allowing market forces to determine production patterns. Although short-term palliatives, such as supply controls, were accepted as useful, and adjustment assistance was considered desirable, the emphasis was on liberalization. Income objectives were to be dealt with in ways that did not distort trade patterns or impose undue costs on other countries. High-level encouragement for this process came from the Tokyo Summit in 1986 and the Venice Summit in 1987. The Toronto Summit in 1988 confirmed the importance of progress in this area of international cooperation, and urged a comprehensive framework for the agricultural talks.

The Uruguay Round

The GATT effort culminated in the launching of the Uruguay Round in September 1986. The Punta del Este Declaration opened up the way for negotiations on all agricultural programs that influence trade in farm products. The section on agriculture called for "greater liberalization" and "more operationally effective GATT rules and disciplines" regarding all "measures affecting import access and export competition." It was made clear that the measures referred to include domestic policies as well as border measures. The inclusion of domestic support policies represented a new departure, which recognized that domestic policy reform provides the key to progress in agricultural trade relations.

The 1987 Proposals

Countries were invited to submit by the end of 1987 their ideas for the conduct of the negotiations, and their proposals for the implementation of the aims of the Punta del Este Declaration. The responses were more far-reaching than most observers expected, and much more imaginative than in previous rounds. It looked almost as if governments were eager to negotiate over policies they had so long held sacrosanct. The U.S. responded first, tabling a proposal in July 1987; the Cairns Group and the European Community followed with their own proposals in October 1987, together with a separate paper by Canada (even though it is a member of the Cairns Group). By early 1988 two more proposals had been introduced, one from Japan and another from the Nordic Countries (Sweden, Finland, Norway, and Iceland). Papers by Korea and by a group of developing countries (Egypt, Jamaica, Mexico, Morocco, and Peru), together with declarations from India, Nigeria, and Switzerland, were also tabled.

The U.S. proposal called for the elimination of "all policies which distort [agricultural] trade" over a ten year period. Certain domestic policies would be exempt, as only having a small impact on production and trade. These would include domestic food assistance programs, international food aid, and any "safety net" farm income programs not linked to the level of production. The Cairns Group shared the U.S. view of a liberalized agricultural market, but added a preliminary stage—a freeze in present subsidy levels. The EC paper appeared more concerned with short-term market imbalances, but agreed that a reduction in support prices was necessary over the long run. At the other extreme, the Japanese position implied that trade problems were the fault of the exporting countries, and that domestic policies were not subject to international negotiation.

Negotiations on domestic policies pose a tricky problem for the comparison of trade effects. The variety of measures used make the process less than transparent. In this respect, the OECD work fitted in well with the GATT aims. All the major proposals, except for Japan's, supported the use of an aggregate measure of support (AMS), based on the OECD's Producer Subsidy Equivalent (PSE), though most countries suggested that this be modified to take into account supply control, and the EC wanted the measure to consider changes in the exchange rate as well.[10] This technique of using an AMS held out the prospect of giving a clear focus to the liberalization aspect of the agricultural talks similar to that of a tariff-cut target in other areas. Countries would be free to modify policies in their own way, subject to the limitation that the overall level of support be reduced according to the agreement.

Discussion of these proposals took place during 1988, in the GATT meetings in Geneva. Seven meetings of the Agricultural Negotiating Group were held between February and November. In addition, the group set up technical working parties to consider the issues of the use of an AMS and of sanitary and phytosanitary regulations. The major participants also spelled out their own proposals over the year in additional papers. The Cairns Group introduced the notion that, in addition to the freeze in support levels, countries should agree to a two-year "down payment" on longer-term reform. This would ensure that policy changes were started during the negotiations, rather than awaiting their conclusion. The EC elaborated on its own short-term proposals, but declined to spell out what they had in mind for the longer-term reform. For much of the year, the U.S. stuck to its position that no short-term action was required, other than an immediate start on the long-term program. By the end of the year, the U.S. indicated its willingness to talk about short-term measures, so long as the goal of long-term elimination of support had been agreed upon. The EC found this position unacceptable, claiming that it went far beyond the mandate of Punta del Este.

Lack of agreement on the longer-term program for agriculture threatened the timetable for the Uruguay Round as a whole. The Midterm Review, held in December 1988 in Montreal, was intended to endorse at the ministerial level the progress so far, and set the agenda for the remaining two years of negotiations. In the event, the Montreal meeting broke up with no agreement on agriculture, and several agreements that had been reached on other issues were put on hold. The GATT director-general was asked to try to find a compromise on this (and some other outstanding issues) in order to allow the round to continue. A compromise framework for negotiations was finally agreed upon on April 7, 1989, in Geneva, to complete the Midterm Review.

The Midterm Agreement

The April agreement did not resolve the many issues that had separated the negotiating positions of the major participants, but it did provide a framework for their eventual resolution. Equally as important, it mandated a timetable for the workplan, which instilled some urgency and discipline into the negotiations. The main provisions of the agreement were a short-term commitment to freeze policy prices, a decision to substantially reduce support levels over a period of years, and an obligation to discuss revision of GATT rules governing agriculture, including those dealing with plant and animal health regulations and food safety. Countries also agreed to specify to what extent developing countries would be expected to accept all the obligations of the GATT rules.

The short-run program called for a two-year freeze in support levels, import access barriers, and national prices (in nominal terms), and registered an "intent" to reduce protection in 1990 by an unspecified amount. Policy prices had been edging downwards since 1986 in the EC, the U.S., and Japan, as a reaction to high budget costs and low world prices at the time. It is not clear whether or not the freeze has had a major impact on this trend. On the other hand, the freeze does appear to have helped guard against an upsurge in domestic prices in response to the higher world prices of the past year.

The second element of the April agreement was long-term reform in agricultural policies. Countries agreed to negotiate a "substantial, progressive reduction in agricultural support sustained over an agreed period of time." The time period remains to be decided, and the level of reductions was not specified. The U.S. appeared to accept that its goal of zero trade-distorting support within ten years was not going to be agreed to by others. Also still to be agreed upon is the mechanism for achieving this reduction. The two main candidates are negotiations on specific policies, including conversion to tariffs, and negotiations of commitments on an aggregate measure of support. It now seems likely that both will be used in various ways during the negotiations.

The third element in the April agreement was for changes in GATT rules. Two rules in particular have been notably significant in agricultural trade. Article XI specifies conditions under which quantitative trade barriers are allowed. Among these exceptions is one for situations where domestic supply is also controlled by quantitative means; some countries would like this exception removed. Participants in the negotiations seem to agree that it is time to get rid of the exceptions and waivers (though the negotiating value of relinquishing a waiver has yet to be tested), and to clarify the status of "grey area" import measures.

Article XVI allows export subsidies on primary products, although stipulating that countries should "seek to avoid their use." Countries have not taken this advice, and even with the negotiation of a Subsidies-Countervail Code in the Tokyo Round, the problem has proved intractable. Moreover, many domestic policies act as indirect subsidies to exports, including U.S. deficiency payments for grains. No consensus has emerged on how to deal with export and domestic subsidies by changing rules. An outright ban on this type of subsidy has been suggested, but the EC views this as both impracticable and inequitable. Constraints on the level of export subsidies may be possible, and the quantities exported with the aid of such subsidies could be limited. With firm world prices, these agreements could be workable. Whether they would be sustainable at a time when world prices were depressed, as happened in 1986, is much less certain.

Less dramatic than support reduction and export subsidies, but still important for agricultural trade, is the area of sanitary and phytosanitary regulations (SPR). The Punta del Este Declaration made special mention of the need to "minimize the adverse effects that sanitary and phytosanitary regulations and barriers can have on trade in agriculture." Behind this apparent agreement, the different approaches to the problem are still to be reconciled. The U.S. view emphasizes the need to base domestic regulations on internationally agreed standards, and to recognize the equivalence of different regulations (i.e., those that provide substantially the same protection to the consumer). The EC has been less explicit on such matters, and has argued for the negotiation of a "framework of rules" governing harmonization of standards. The need for equivalence is not fully accepted, and the EC also places less emphasis on the scientific basis for international regulation. The Cairns Group also favors a negotiation on a long-term framework based on strict justification for protecting human and animal health. It recognizes the value of equivalence where harmonization is not possible, but the unwillingness of Australia to sign the Standards Code has left it difficult for the Cairns Group to support the U.S. position more fully.

The Comprehensive Proposals

Under the timetable of the Midterm Agreement, countries were to present their ideas for a package of reforms, including the reduction in support and the changes in GATT rules, by the end of 1989. In October, the U.S. presented the first of these "comprehensive" proposals, which builds upon earlier papers but contains considerably more detail on the implementation of reforms. The proposal itself is organized by category of policy instrument–import access, export subsidies, domestic subsidies, sanitary and phytosanitary barriers, and distinctive treatment for developing countries. As discussed above, the three negotiating methods under consideration are changing GATT rules, reducing support by an AMS, and negotiating on specific policies and commodities.

The U.S. comprehensive proposal places a heavy burden on negotiating changes in GATT rules. Import access is to be tackled by getting rid of all nontariff import barriers. Tariff quotas would protect export interests over a ten-year transition period, and other safeguards are designed to prevent too rapid an increase in imports. Specific levels of tariffs, along with the amount of the tariff-rate quota, would be negotiated. Quantitative restrictions would be removed by the end of the transition, thus eliminating the need for the current waivers and exceptions. Export subsidies would also be removed (over a five-year period) and the GATT rules changed accordingly.[11] For domestic subsidies, a set of regulations would define those that are prohibited (i.e., those that reward farmers on the basis of current output) and those that are permitted (i.e., those that are effectively decoupled or only indirectly linked to output). Domestic subsidies falling between these two categories would be "disciplined" by negotiated reductions in the level of support afforded by such policies. New procedures would be developed to strengthen notification, consultation, and dispute-settlement obligations for sanitary and phytosanitary trade barriers.

Countries would need to present and agree to sweeping changes in domestic price-support programs, each being locked in to a ten-year program of domestic policy reinstrumentation. The U.S. originally pushed for the use of an AMS, as a way of capturing the totality of agricultural policy instruments in one comparative figure. When it became clear that a zero target for the protection level was not feasible, concern grew that countries might try to evade the trade implications of support reductions by shifting policies. It was feared that reductions in the AMS might not be translated into visible trade gains. Though many of the these concerns can be met by the appropriate definition of the AMS, the feeling grew that it would be desirable to deal with specific policies to enable exporters to target priority importer practices. The AMS finds only a minor role in the U.S. proposal, as a way of measuring these policies that fall neither into the prohibited nor the permitted list. These policies, largely input and investment subsidies that are not tied to particular commodities, would be subject to a phase-down over a period of years.

The Cairns Group tabled its own comprehensive proposals for the agricultural negotiations in December 1989. Like the U.S. plan, the Cairns Group emphasized import access and the control of export subsidies. Members also advocated conversion of nontariff measures to tariffs, and stressed the need for bringing all import measures under GATT coverage. For export subsidies, the Cairns Group proposal argued for a freeze and a progressive phase-out of these programs. They also suggested that the "option" of converting export subsidies to food aid should be made less attractive by limiting this type of aid to outright grants of food. The Cairns Group sees a somewhat larger role for the AMS than does the U.S., but treats it essentially as a scorecard to keep track of progress obtained by specific bargaining. The Cairns Group proposal on the reform of domestic policies is slightly more flexible than that of the U.S., allowing somewhat more choice of domestic policy instruments.

The EC "global proposal" showed perhaps the most willingness to bend from their previous position.[12] They reluctantly conceded that tariffication, in a modified form, could replace the present levy system on EC imports, and a careful reading between the lines suggests that a similar modification could be applied to EC export subsidies. So long as equivalent concessions were available on the main policy instruments of other trading partners, one could envisage a significant modification to EC policy following from such an agreement. The main thrust of the EC proposal is not, however, on changes in instruments. To the EC, negotiations on specific policies means attacking the CAP; by contrast, AMS negotiations would move some of the focus back onto the U.S. and Canada, who proved to have surprisingly high PSEs in 1986 and 1987. And an AMS negotiation might allow the EC to spare some sectors, and even to achieve an element of "rebalancing" in its policy. The EC is now the main enthusiast for an AMS approach.[13] Whether other countries will shift their own positions to accommodate the EC on this issue remains to be seen.[14]

Impacts of GATT on National Policies

The changes in national farm policies needed to improve both the domestic performance and the international acceptability of farm programs will not be easy. They will require significant skill in international negotiations and considerable political will at home. "Success," in some countries, requires that traditional forces that opposed these changes, and that have been able to block them in the past, be persuaded to remove their opposition. The Uruguay Round cannot dictate domestic farm policy change; any attempt to do so would condemn it to failure. But the round cannot ignore domestic policies and concentrate solely on trade rules or the exchange of trade concessions. Such tinkering with the symptoms has been a major part of the problem in the past. What is now possible is an outcome that *assists* the process of domestic reform, just as the trade problems themselves will be reduced by such reform.

It now appears possible that the Uruguay Round negotiations on agriculture could end in agreement which would:

a) specify a substantial reduction in the levels of support under farm programs;
b) encourage the reinstrumentation of policies toward more decoupling in domestic subsidies;
c) promote some recoupling of domestic markets with world market conditions (such as through tariffication);
d) restrain export subsidies, even if not removing them altogether; and
e) modify GATT rules to encourage and underpin these developments and to ease the resolution of any conflicts that may arise in the future.

Observers of the process tend to fall into two camps. One group, who might be called the "realists," consider the agricultural negotiations in the

Uruguay Round largely an exercise in illusion, a game played by trade officials at the periphery of agricultural policy. Whatever comes out of the round, they feel, will amount to minor changes in the trade rules and some collective but empty encouragement to domestic policy reforms. The chances of major farm policy changes being agreed upon in Geneva are remote; to pretend otherwise is fanciful at best and possibly even dangerous. It flies against all the experience of recent history and ignores political realities. Under such a "realistic" interpretation, the U.S. and the EC should not look to the GATT to reduce trade tensions. Some hard bargaining on a bilateral basis could yield results, but essentially the solutions lie in domestic politics and initiatives.

History and political logic is so firmly on the side of this interpretation that it seems foolhardy to argue otherwise. But there is another more "optimistic" interpretation of current events that is worth passing consideration. Suppose that special-interest politics and middle-class entitlement programs have now become less attractive to politicians. Imagine agricultural price policy becoming identified with transfers to wealthy families and with the degradation of the environment. Add a sense of frustration over agricultural trade squabbles from those in charge of commerce and commercial policy. It is not inconceivable that the GATT round under these circumstances could play a nontraditional role. The tail of trade policy could for once wag the domestic-policy dog. A bold agreement in the GATT could essentially dictate the framework for domestic policies for the next decade.

If the realist interpretation is proved correct, then the unimportance of the GATT in agricultural trade will have been confirmed. Once again, negotiations will have failed to find the elusive formula that allows domestic policies to run under internationally agreed rules, rather than bending international rules to fit domestic policies. There may be enough of an agreement to allow the Uruguay Round to be counted a "success" without, at the same time, challenging the dominance of domestic policies. Some deals could be struck that promised exporter constituencies greater access without seriously constraining domestic support. In such a case, the underlying problems will remain so long as the instrumentation of those policies, and the general level of support, is unchanged. Domestic pressures will work slowly to move support away from surplus-generating measures, but without external disciplines the temptation to resort to passing the burden to other countries will be irresistible. If world market demand were to become weak, then all the problems of the early 1980s will return. A scramble for markets will push up program costs and set in motion another cycle of reform.

The optimistic scenario offers a short cut to this process. By constraining domestic policies within effective GATT rules, countries would incur much greater risks when running policies that offer excessive protection to agriculture. Domestic policies would run up against these constraints and adapt more quickly. Trade conflicts would not disappear overnight, but they would take place in a framework of rules and obligations. This attractive scenario could just be enough to persuade negotiators to defy history and challenge political reality. If that were the case we could look back to the Uruguay Round as a turning point in international cooperation. The GATT would then truly have a role to play in agricultural policies and markets.

Notes

1. The establishment of the GATT and its legal constitution is described in McGovern (1986).
2. For a fuller discussion of the treatment of agriculture in the GATT see Hathaway (1987) and Hartwig, Josling, and Tangermann (1989).
3. The waiver allowed quantitative import restrictions, under Article 22 of the Agricultural Adjustment Act, to be used even in the absence of domestic supply control.
4. This situation may now be changing. Recent panel rulings on US sugar, EC oilseeds and Canadian dairy policies, each appear to have inescapable implications for a change in domestic policy.
5. It is ironic that the most notable agricultural "achievement" of the GATT, to bind EC duties on oilseeds and cereal substitutes, came in an otherwise unsuccessful round of negotiations, which ended with little agreement on liberalization of trade.
6. In the nonagricultural area, the Kennedy Round proved a significant success, with an across-the-board cut of 50 percent in bound tariffs.
7. For a discussion of the experience with Article XVI and the Subsidies Code, see Hartwig, Josling, and Tangermann (1989).
8. OECD (1987).
9. For a fuller discussion of the OECD work see Sanderson, Warley, and Josling (1990).
10. For a discussion of the use of an AMS in trade negotiations, see Hartwig, Josling, and Tangermann (1989).
11. The U.S. proposal also calls for the removal of the clause that allows export restrictions in time of domestic shortage.
12. Tangermann (1990) makes the point that the U.S. has considerably hardened its line on the reinstrumentation of domestic policies, while abandoning its goal of a zero level of trade-distorting support. The EC position has moved some way to meeting U.S. requests on changes in domestic policy, perhaps more than the rhetoric of the proposal would suggest. However, the EC also appears to have made such policy changes contingent upon agreement to "rebalancing" protection among sectors.

13. The EC has labeled its measure a Support Measurement Unit (SMU). It is essentially the OECD PSE corrected for supply control and measured using reference prices and exchange rates rather than current world prices.

14. The extent of the "common ground" between the proposals is explained in Bredahl, et al. (1990).

References

Bredahl, M., et al. *Report of the Task Force on the Comprehensive Proposals for Negotiations in Agriculture.* IATRC Working Paper No. 90-3 (1990).

Cohn, Theodore H. *The International Politics of Agricultural Trade.* Vancouver: University of British Columbia Press, 1990.

GATT. *Submission of the United States on Comprehensive Long-Term Agricultural Reforms.* Geneva: GATT, 1989.

———. *Comprehensive Proposal for the Long-Term Reform of Agricultural Trade: Submission by the Cairns Group.* Geneva: GATT, 1989.

———. *Global Proposal of the European Community on the Long-Term Objectives for the Multilateral Negotiation on Agricultural Questions.* Geneva: GATT, 1989.

———. *Negotiating Group on Agriculture: Submission by Japan.* Geneva: GATT, 1989.

———. *Ministerial Declaration on the Uruguay Round.* Geneva: GATT, 1986.

———. *United States Proposal for Negotiations on Agriculture.* Geneva: GATT, 1987.

———. *European Communities Proposal for Multilateral Trade Negotiations on Agriculture.* Geneva: GATT, 1987.

———. *Japanese Proposal for Negotiations on Agriculture.* Geneva: GATT, 1987.

———. *Proposal of the Nordic Countries.* Geneva: GATT, 1987.

———. *Final Text: Mid-Term GATT Review.* Geneva: GATT, 1989.

Hartwig, Bettina, Timothy Josling, and Stefan Tangermann. *Design of New Rules for Agriculture in the GATT.* Report to NCAP, Washington, D.C.: Resources for the Future, 1989.

Hathaway, Dale E. *Agriculture and the GATT: Rewriting the Rules.* Washington, D.C.: Institute for International Economics, 1987.

Josling, Timothy E. *Agriculture in the Tokyo Round Negotiations.* London: Trade Policy Research Centre, 1977.

McCalla, Alex. "The History of Agricultural Protectionism." *Journal of Agricultural History* (April 1971).

McGovern, Edmond. *International Trade Regulation: GATT, the United States and the European Community.* Exeter: Globefield Press, 1986.

Moyer, H.Wayne and Timothy Josling. *Agricultural Policy Reforms: Politics and Process in the EC and the USA.* Ames: Iowa State University Press, 1990.

OECD. *National Policies and Agricultural Trade.* Paris: OECD, 1987.

Sanderson, Fred H., T.K. Warley, and Tim Josling. "The Future of International Agricultural Relations: Issues in the GATT Negotiations." In Fred H. Sanderson, ed. *Agricultural Protectionism in the Industrialized World.* Washington, D.C.: Resources for the Future, 1990.

Tangermann, Stefan. "Options and Prospects: A Feasible Package." Department of Agricultural Economics and Business, University of Guelph, 1990.

Warley, T.K. "Implications for Canadian Agrifood." Department of Agricultural Economics and Business, University of Guelph, 1990.

———. "Western Trade in Agricultural Products." In Andrew Shonfield, ed. *International Economic Relations of the Western World 1959-71: Politics and Trade.* London: Oxford University Press/RIIA, 1976.

SECTION THREE

Prospects for a New World Agricultural Order

7

Prospects for the Uruguay Round in Agriculture

C. Ford Runge

The Uruguay Round of Multilateral Trade Negotiations (MTN) is scheduled to conclude in December 1990. Whether it will be regarded as a successful round of talks depends in part on the negotiations in agriculture, which have proven as difficult as many predicted. This chapter considers three rather different aspects of the agricultural negotiations. First, what is the "core" of agreement most likely to emerge in agriculture, especially between the major antagonists, the U.S. and the European Community? Second, how will the domestic politics of the 1990 Farm Bill interact with the negotiating process? And third, in what ways will the results of the Uruguay Round condition subsequent rounds of MTNs under the General Agreement on Tariffs and Trade?

In brief, I will argue first that a "framework agreement" in agriculture will emerge, which closely resembles the conception put forward in the most recent proposals of the U.S. and Cairns Group. Whether this framework results in substantive reductions in levels of protection depends, at its core, on the willingness of the U.S. to reduce and eventually to eliminate export subsidies in return for substantial reductions in the level of export restitutions paid by the EC. While difficult to achieve, some experts are currently optimistic about the chances, even for "tariffication" and "rebalancing." Second, the 1990 Farm Bill will produce incremental movements in the direction of "decoupling," cloaked in the language of "flexibility," regardless of the outcome of the Uruguay Round. These modest reforms will reinforce the U.S. position in GATT. However, more substantive movements in the direction of tariffication are likely to be opposed by the domestic sugar, dairy, and peanut lobbies, in spite of some evidence that they might gain from a conversion of quotas to tariffs. In addition, Congress will be likely to demand authorization of a "war chest" of retaliatory measures in case the GATT talks proceed unsatisfactorily. Third, the Uruguay Round has ushered in a new complex of nontariff trade barriers (NTBs) that are likely to grow in importance in future trade negotiations. These new NTBs involve the use of health, safety, and environmental regulations as effective barriers to trade—what I call "ecoprotectionism." Several recent cases, including the hormones dispute between the U.S. and EC, and a more recent U.S./Canada dispute over landing requirements for Pacific Coast salmon and herring, provide glimpses of this growing threat to liberal international trade.

Framework Agreement and What Else Besides?

In the latter part of 1989, the U.S. put forward its final proposal for agriculture in GATT. In brief, the U.S. proposal calls for the elimination of export subsidies over a five-year period, the phase-out of trade-distorting domestic programs, and the conversion of nontariff barriers, such as quotas, to tariffs, known as tariffication.

It may be useful to summarize in graphical form the conception underlying the U.S. proposal (see Runge, 1990, 1988). In general, trade-distorting measures may be thought of in terms of their effect on (a) exports, (b) imports, and (c) output. Tariff equivalents describe distortions on the import side, and subsidies on the export side. Output distortions resulting from internal policies are derived with respect to their effect on production from internal policies are derived with respect to their effect on production in domestic markets. In each case, policies may either promote or retard exports, imports, or output.

With respect to exports, a policy has a distorting trade effect if either buyers or sellers in the domestic market face different conditions from those who participate in the cross-border market. Such a definition encompasses not only policies that affect the difference between export and domestic prices, such as export taxes and subsidies, but also nonprice protective barriers, such as voluntary export restraints. As shown in Figure 7.1, such policies may distort trade either by artificially promoting exports (as in the case of the U.S. Export Enhancement Program), or by artificially retarding them (as in the case of Argentine export taxes or various countries' voluntary export restraints). Over the remainder of the Uruguay Round, the attempt will be to define and to set GATT-negotiate limits, for each country, on three types of policies: those that are definitely slated for elimination, preferably as soon as possible ("Red Light" policies); those that may remain

Figure 7.1 Export-Distorting Policies

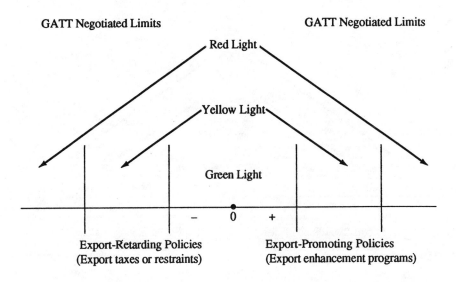

in place in the short-run, but are to be modified and reformed during a transition phase ("Yellow Light" policies); and those that are sufficiently nondistorting to remain in place indefinitely ("Green Light" policies).

Similarly with respect to imports, a policy has a distorting trade effect if either buyers or sellers in the domestic market face different conditions from those who participate in the cross-border market. As shown in Figure 7.2, policies that retard imports, such as quotas, explicit tariffs, or health, safety, and other sanitary or phytosanitary restrictions, are one side of such distortions. On the other side (less frequently mentioned) are policies that artificially promote imports. An example might be environmental regulations on fruit and vegetable production that prohibit the use of certain cost-saving chemicals in the U.S., leading to incentives to import foreign fruit and vegetables whose production complies with such practices. Because domestic growers quickly realize the effect of these regulations, calls for import protection through health and safety standards applied equally to foreign produce are soon heard, converting the regulations from import-promoting to import-retarding (see Runge and Nolan, 1990). In principle, either type of distortion can be expressed as a tariff equivalent, with import-promoting policies defined as a negative tariff. Once again, the issue is which policies are determined to be definitely out-of-bounds ("Red

Figure 7.2 Import-Distorting Policies

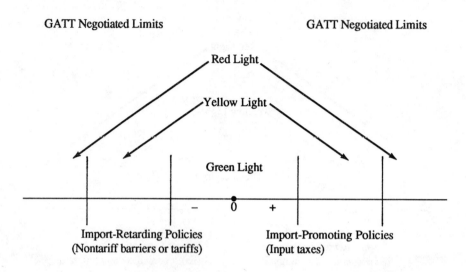

Light"), which are undesirable and to be phased out over time ("Yellow Light"), and which are acceptable ("Green Light").

Figure 7.3 deals with policies that have an effect on domestic production. As shown, these policies may be negative, such as U.S. and European set-aside programs that pay farmers not to produce; or they may be positive, such as price supports tied to specific crop yields and acres of production.

The goal of U.S. domestic agricultural policy in the Bush administration is generally to eliminate policies that are most distortive of production decisions ("Red Light" policies), including large set-asides and high price supports, and to phase out those that have tended to distort production over time ("Yellow Light" policies), such as crop-specific acreage bases. What remains ("Green Light" policies) will be programs in which farmers are relatively free to plan whatever crops are most in market demand, with support paid not to specific crops, but on the basis of some type of income criteria.

Overall, progress in the present GATT negotiations can be defined in terms of this framework. And this development depends on an agreement to eliminate, according to a well-defined timetable, a specific set of "Red Light" policies in each realm (exports, imports, and output), and to designate a set of "Yellow Light" policies for discussion in subsequent years. It seems inevitable that successful negotiations will ultimately involve agreements to end specific *policies,* and that such political decisions

Figure 7.3 Output-Distorting Policies

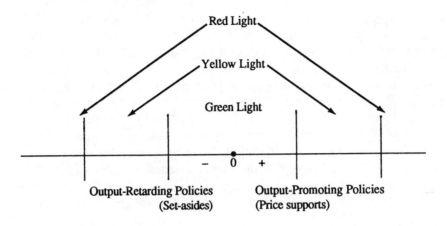

cannot be finessed by an agreement simply to achieve an aggregate level of support or level of tariff or subsidy. This is the route sometimes suggested by advocates of a single aggregate measure, such as the Producer Subsidy Equivalent (PSE). As Hertel (1989a, b) has recently shown, a given reduction in the aggregate level of support can be achieved with a myriad of options, many of which have widely varying effects on exports, imports, and output. His analysis shows that aggregate measures, because they abstract from this complexity, "underidentify" the problem, and thus do not provide sufficient discipline to achieve long-lasting reform. The PSE measure, whatever its virtues as an analytical device, cannot be a substitute for the hard political choices that accompany a negotiation.

How likely is progress, given the proposals of the major negotiating countries. In November 1989, the Cairns Group of fourteen agricultural exporters put forward a close cousin of the U.S. scheme. Reflecting the composition of the Cairns Group, the major difference was in the attention paid to the interests of developing countries, allowing them greater flexibility in implementing reforms. In the context of the framework described above, what is red or yellow light is defined with more leeway for developing countries.

The EC proposal has been interpreted in the U.S. as continuing the European tradition of conceding little and demanding much in return. Close observers of the Brussels scene, however, see more room for compromise than meets the eye. In particular, the EC proposes to give ground on the possibility of tariffication. In a recent paper, Stefan Tangermann (1990) argues that due in part to the unfavorable ruling on the EC oilseed regime rendered by a GATT panel in December 1989, the Community is prepared to make concessions which, in essence, reduce border protection through tariffication in combination with "rebalancing." What makes U.S. negotiators suspicious is that the newfound EC commitment to tariffication is vague and is even contradicted in their proposal, while the commitment to rebalancing is hard and fast.

Rebalancing means that increases in protection at the border for oilseeds (currently excepted under the "zero duty binding") would be traded for reduced levels of protection for grains. All of this should be measured, according to the EC proposal, in "support measurement units." While conceding some room for tariffication, if tied to rebalancing, the EC also demands that U.S. deficiency payments be converted to tariffs, and that EC internal supply management be retained. While cynics might see little hope that the Community is ready to compromise, Tangermann is remarkably upbeat.

> The significance of this latest move of the EC can hardly be overestimated. In essence it means that the EC is now willing to consider fundamental changes to the way in which it operates its agricultural market regimes. In particular, the variable levy system as such is no longer sacrosanct. The EC has thus left its long held negotiating corner and moved a very considerable step towards the center of the negotiating positions of the different parties (p. 5).

Before getting caught up in this enthusiasm, it is chastening to note (as does Tangermann) that the proposal also states that "Basing protection exclusively on customs tariffs and envisaging, after a transitional period, the reduction of these tariffs to zero or a very low level, would lead to trade in agricultural products on a totally free and chaotic basis."

What the EC means is thus ambiguous, but can be interpreted as support for *some* tariffication, less than total. Tangermann goes on to propose a modified tariff at time (t), MT (t), of form:

$$MT(t) = FC(t) + \alpha \, DIF(t)$$

where FC (t) is the "fixed" component of the tariff in year (t) and DIF (t) is the "floating" or differential component, which moves with world market prices, as the variable levy does now. The parameter α $(0 \leq \alpha \leq 1)$ determines the "weight" given to the floating component, which would become the focus of negotiation, and might be set at 0.7 at the beginning of an adjustment period and perhaps 0.3 toward the end, moving closer to a fixed tariff ($\alpha = 0$) over time. Such a framework could provide a mechanism for fixed versus floating export subsidies as well, which could then be walked down according to a schedule, consistent with red and yellow light designations.

Tangermann argues that the EC's loss on the GATT oilseeds panel strengthens the internal case in the Community for converting to tariffication in this particular sector as an alternative to current price supports. If so, this opens the possibility of a similar concession on the part of the U.S. in its sugar quota regime, which was found unacceptable in an earlier 1989 GATT panel. In sum, the EC and U.S. have incentives to go down the tariffication road together, designating the current EC oilseeds and U.S. sugar policies as "Red Light," and working toward "modified tariffication" in tandem.

A final negotiating player is Japan, which continues to let the EC do much of its bidding, while gladly calling for the elimination of export subsidies, since its agricultural sector exports so little. At the center of the Japanese proposal is "food security," which is used to justify the continuation of border measures, exceedingly high internal price supports, and a variety of nontariff barriers.

Given the continued divergence of views, what is likely to emerge by the end of 1990?

It is possible that consensus will be reached on a framework agreement, so long as it commits the negotiating parties to little actual reform. To *agree to designate* policies as "red light," even to agree to modified tariffication, need not involve more than a commitment in principle. This is not the same thing as agricultural policy reform, which is ultimately a domestic political issue. The EC can continue its own policies while agreeing in principle to sort them out over time. Similarly, Japan can use food security as a blanket exemption for its own most egregious distortions, while agreeing in principle that, in some future period, reform will be necessary along the lines of the proposed framework. This agreement to a mere framework may seem a hollow victory, but it can be regarded as a useful first step. It would not be surprising to see various contracting parties go somewhat further, by each offering to sacrifice some aspect of their own policies as an example of a "Red Light" distortion, primarily because they would like to be rid of it for internal political reasons. This is the case in the soybeans-for-sugar trade concessions mentioned above.

Assuming such a framework can be reached, and that token sacrifices might also be made, is there any real prospect of substantive reforms? It would not be advisable to bet on it, but if it does occur, the core of the agreement must be a decision on the part of the U.S. to reduce and eventually eliminate the Export Enhancement Program (EEP) and to reduce domestic levels of support, in return for major European reductions in export restitutions, as well as changes in the variable levy. On the U.S. side, the EEP has been something of a dog in financial terms, despite the enthusiasm shown for it by beneficiaries, such as wheat growers. Coughlin and Carraro (1988) noted that the cost of EEP subsidies for wheat averaged $4.08 per bushel, compared with an average market price at the Gulf of $3.16, implying that it would have been cheaper to destroy the wheat on the farm and pay farmers the difference, than to ship it halfway around the world at subsidy.

It would be nice to sacrifice EEP and to reduce domestic supports purely for domestic budgetary reasons. Conjecture remains, however, as to what Europe might sacrifice. Here, again, the key is the rebalancing question. European sacrifices of export restitutions in grains are possible, *if* oilseeds can be given a measure of protection, perhaps through modified tariffication. It is doubtful that such protection will be too high, lest the EC livestock industry protests the loss of cheap U.S. feed. However, will the U.S. soybean industry concede the need to "close the CAP" further, losing the zero duty binding, in return for gains in the maize and wheat markets? In the Midwest, the answer may be yes, but the South's soybean growers have fewer options. In sum, a core agreement between the U.S. and EC will be difficult to achieve unless domestic interests provide the requisite political support.

The Impact of the 1990 Farm Bill Debate

These domestic interests are now circling the wagons for the 1990 Farm Bill debate. Many commodity groups, and their elected representatives in Congress, seem unable to decide if GATT is an ineffectual group of diplomatic Dr. Jekylls, or a threat to farmers, capable of becoming a monstrous Mr. Hyde. The Congress bravely asserts that it does not believe in monsters, yet keeps peeking out from under its bed covers, just to make sure. On the one hand, one hears that "GATT is dead," and that the U.S. must not continue to pay lip service to an outmoded and toothless institution. This point of view is held not only by xenophobes and "America Firsters," but also by such respected economists as Lester Thurow, who is neither (see Litan and Suchman, 1990). On the other hand, one hears that "We will not allow the Farm Bill to be written in Geneva." Apart from the fact that the U.S. negotiators (much less their foreign counterparts) would be unlikely voluntarily to come within miles of farm legislation, this concern suggests the fear that GATT may have teeth after all.

The domestic farm legislation passed in 1990 will contain changes that support the U.S. position in GATT. This support will come in positive reforms that move modestly toward decoupling, and in negative threats of retaliation if GATT is not a "success." On the positive side, the move toward flexibility is now widely popular. Flexibility means that farmers will be given greater leeway to plant a range of crops without losing cropping "bases" on which government program payments are made. The administration proposal advocates return to a Normal Crop Acreage or "whole farm" base, in which current bases for various crops would be merged into one single accounting unit. This approach has broad support, concentrated especially in commodity groups like the soybean, oats, and barley growers, who argue persuasively that soybeans, oats, and barley are "crowded out" due to high relative deficiency payments paid to corn and wheat.

Two recent reports illustrate the reasoning behind increased flexibility, which has trade, farm income, and environmental components. The trade argument comes especially from the U.S. soybean sector, which, after being the predominant world supplier form the 1950s through the 1970s, lost substantial market share in the 1980s. This loss is traceable in part to disincentives to grow soybeans at the farm level, and in part to EC subsidies. Figure 7.4 illustrates oilseed planting trends, while Figure 7.5 shows soybean and soybean meal exports. Abel, Daft, and Early (1990) summarize the domestic farm income distortions resulting from current program.

> During the 1986-89 period, target prices for corn, sorghum, rice, wheat, and cotton were comparatively high relative to various measures of production costs. At the other extreme, average target prices for barley and oats were relatively low and oilseeds did not have target prices. These distortions encouraged production of those crops that were most profitable when government payments are included and discouraged production of less profitable crops, namely those with low or no target prices. There is also a higher degree of certainty about revenue from target price crops and this was important to some producers and farm lenders during the mid-decade farm financial crisis (pp. 12-13).

The final justification for more flexible plantings is an environmental consideration (Runge et al., 1990). By encouraging more crop rotation and green manures, less restrictive base acreage requirements would be likely to lower the repetitive cropping of highly erosive crops that also have high nutrient and pesticide demands.

In a comprehensive review of such a flexible program, the Congressional Budget Office (CBO, 1989) estimates that farm incomes would fall slightly

Figure 7.4 Oilseed Planting Trends

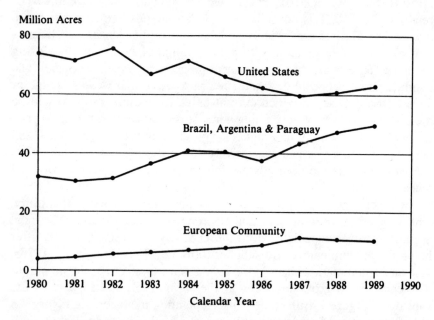

(as would government program costs), but that in return, farmers would be allowed much greater freedom to pursue marketing opportunities. If program costs are to be cut anyway for broader budgetary reasons, "flexibility" affords an attractive way to do it, and is, in effect, an incremental step toward decoupling, and thus trade-liberalizing overall.

In contrast to flexibility, the idea of tariffication has not gone down well in Congress, for several reasons. Most obviously, it is an unknown alternative to the well-known quantitative import restrictions applied in the U.S. sugar, dairy, peanut, and tobacco programs, which have served producers well. At the political level, the House and Senate agriculture committees fear that conversion of quotas to tariffs would shift sovereignty over agricultural prices to the trade subcommittees, such as that of the House Ways and Means Committee, where sympathy for consumers far outweighs that for farmers. From a regional perspective, the commodities most affected by tariffication are concentrated in the South and upper Midwest, where enthusiasm for trade liberalization is less than on the coasts. Combining tariffication with the opposition to rebalancing exercised by southern soybean interests adds up to a powerful southern bloc likely to

Figure 7.5 U.S. Soybean Exports and U.S. Soybean Meal Exports

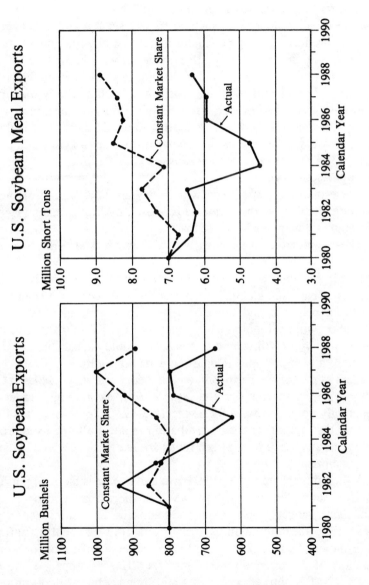

oppose a substantive compromise between the U.S. and EC that includes too much tariffication or rebalancing.

Ironically, there is reason to believe that at least some of these U.S. producers might actually be better off under tariffs than quotas, assuming that the fixed component of such tariffs was set at high levels to begin with and only reduced through formulas determined by multilateral agreement, which would proceed slowly. Beet sugar growers in the Red River Valley of the Dakotas and Minnesota, for example, are low-cost producers who are increasingly undercut by sugar processors, such as candy manufacturers, who have made use of free trade zones or foreign subsidiaries to avoid using high-priced U.S. sugar. A tariff wall would more effectively insulate these growers, while also producing tariff revenues and rewarding their low costs. The primary fear of beet growers is that admitting the feasibility of tariff protection will set them on a slippery slope toward freer trade, which is worse than the slope down which they are now sliding. However, an agreement based on the formulas proposed by Tangermann (1990), cited above, might help to allay these fears.

These concerns have a negative side, which is also likely to enter the Farm Bill in the form of authorized retaliation if the GATT talks do not succeed; this approach is the same as that which characterized the use of EEP as a "bargaining chip." What counts as success, like beauty, is in the eye of the beholder, and how Congress defines the trigger that sets off such retaliation will be important. Opponents of GATT will want a hair-trigger, while supporters will want a soft one. It would appear that the U.S. trade representatives and U.S. Department of Agriculture will concede such triggers as part of the Farm Bill, on the theory that they will keep Congress busy doing something it likes (posturing as a tough trader), while apparently upping the ante for a successful GATT outcome. It is important to note that authorizing such retaliation is not the same as appropriating money to undertake it, and such funds are likely to be scarce in the current budget climate.

A final dimension in the 1990 Farm Bill debate is the rising influence of environmental interest groups. While there may appear to be little connection between these groups and trade talks on first inspection, it seems clear that the increasingly important role of health, safety, and environmental concerns may well be of larger significance for world trade than any of the issues discussed thus far. The full impact of these trends will not be felt until after the Uruguay Round. But if import protection in traditional agricultural programs can be seriously addressed, a new form is likely to take its place, one which is inherently much more difficult to isolate and thus to "tariffy." This market access issue may be termed the rise of "ecoprotectionism."

Ecoprotectionism and Future Trade Negotiations

This chapter concludes with a brief discussion of ecoprotectionism, by which is meant the use of health, safety, and environmental regulations as effective barriers to trade (see Runge and Nolan, 1990).

Ecoprotectionism involves the internationalization of issues related to food, health, and safety. This gives rise to problems created by national income disparities and the differing priorities of national governments. A recent U.S./Canada dispute over salmon and herring exports provides legal precedents for future international actions, which will involve GATT in a multi-tiered set of international standards that can help distinguish legitimate health and environmental regulations from disguised nontariff barriers.

On January 1, 1989, the European Community announced a ban on all beef imports from the United States containing hormones used to help increase cattle growth. City health risks, the EC action touched off a cycle of retaliation worth hundreds of millions of dollars that has affected the world trading system, and is still being negotiated. This apparently isolated example of health regulations acting as trade barriers is part of an emerging pattern of environmental and health issues that will have major consequences for the world economy.

In September 1989, the European Commission took up discussions of further rules to restrict imports of cattle or dairy products produced with the bovine growth hormone BST (bovine somatotropin). BST is also at the center of domestic controversies over the safety of food supplies in the U.S. and Canada.

Senator Pete Wilson (R-CA), a candidate for governor of California, introduced federal legislation in December 1989 that would ban companies from exporting pesticides that are illegal in the U.S. Responding to the Western Growers Association, Wilson stated that "export of dangerous pesticides creates a competitive inequity between foreign and American farmers and growers." A spokesperson for the growers argued that "we are under extreme pressure from foreign farmers," noting hundreds of growers who have gone out of business because of competition with Mexico and other countries where "they can use whatever [chemicals] they want in most cases."

In October 1989 a dispute settlement panel formed under the U.S./Canada Free Trade Agreement (FTA) determined that Canadian restrictions on foreign salmon and herring fishing constituted an effective barrier to trade, despite the fact that Canada justified them as environmentally motivated conservation measures under Article XX of GATT.

These examples are part of an emerging pattern in which environmental and health risks are increasingly traded among nations, along with goods and services. These risks are the opposite of services; they are environmental and health disservices traded across national borders. They arise directly from the transfer of technology, and will increasingly affect international investment flows, trade and development, and the relative competitiveness of national industries and agriculture.

This pattern of trade underscores the problem of formulating government policies in an interdependent world economy. While the United States and other signatories to GATT pursue more open borders in the ongoing Uruguay Round of trade negotiations, the role of national health, safety, and environmental regulations grows in importance for domestic electorates, especially in the wealthy countries of the north. Increasingly, different national regularity priorities will pose problems for trade harmonization, blurring the distinction between domestic and foreign economic policy. Without additional attempts to come to terms with environmental issues through multilateral institutions such as GATT, differences in national regulatory approaches will bedevil both the environment and the trade system in the next decade and beyond.

The examples cited above demonstrate that environmental regulations are not purely domestic policy issues. Indeed, there has been long-standing recognition of the possibility for conflicts between national environmental policy and more liberal international trade. The GATT articles, adopted by the contracting parties in 1947, explicitly recognize the possibility that domestic health, safety, and environmental policies might override general attempts to lower trade barriers (Jackson, 1969). GATT Article XI, headed "General Elimination of Quantitative Restrictions," states in paragraph (1):

> No prohibitions or restrictions other than duties, taxes or other changes, whether made effective through quotas, import or export licenses or other measures, shall be instituted or maintained by any contracting party on the importation of any product of the territory of any contracting party or on the exportation or sale for export of any product destined for the territory of any other contracting party.

Yet Article XX, headed "General Exceptions," provides

> ... nothing in the Agreement shall be construed to prevent the adoption or enforcement by any contracting party of measures:

> ... (g) relating to the conservation of exhaustible natural resources if such measures are made effective in conjunction with restrictions on domestic production or consumption;

provided that such measures:

> ... are not applied in a manner which would constitute a means of arbitrary or unjustifiable discrimination between countries where the same conditions prevail, or a disguised restriction on international trade.

A similar set of exceptions was applied to health-related measures under Article XX(b). GATT law emphasizes that any restrictions imposed on foreign practices for environmental or health reasons must also reflect a domestic commitment, so that the exception cannot be misused as a disguised form of protection.

Despite substantial attention to both technical standards and nontariff barriers in the Uruguay Round, it is still unclear when and where such standards constitute an unnecessary obstacle to international trade. Although a GATT technical working group on sanitary and phytosanitary measures continues its work, the temptation to use environmental and health standards to deny access to home markets is, if anything, stronger now than in the 1980s. As the European Community moves toward its goal of market integration in 1992, it will have strong incentives to create common regulations for internal purposes, but also to impose restrictions vis-a-vis the rest of the world. A similar propensity may occur as a result of harmonization under the U.S./Canada Free Trade Agreement. Even if *national* standards can be harmonized, moreover, there is every reason to expect subnational jurisdictions to utilize various health and environmental standards to protect certain markets.

Underlying the development of these trade tensions are fundamental differences in the views of developed and developing countries (the "North" and "South") concerning the appropriate level and extent of environmental regulation. Differences in the domestic policy response to these problems are well represented in the food systems of the North and South. Since so much recent attention has focused on food and agricultural chemical use in the North, and because the agricultural sector is of key importance in almost all developing economies of the South, it provides a useful case in point.

In the developed countries of North America and Western Europe, the food problem arises not from too little food and land in production, but rather from too much. As predicted by Engels' Law, the incomes of developed countries have increased, and the share of this income spent on food has fallen in proportion to other goods and services, making food an "inferior good" in economics jargon. In contrast, environmental quality and health concerns have grown in importance with increasing income levels, becoming what economists call "superior goods," in the sense that they play a larger role in the national budget as national incomes increase (see Runge, 1987).

The implications of these trends with regard to competition are not lost on northern producers, who have been quick to see the trade relevance of environmental and health standards. Growing consumer concerns with the health and environmental impacts of agriculture create a natural (and much larger) constituency for nontariff barriers to trade, justified in the name of health and safety. As between countries in the South, obvious differences in values also exist, although the regulatory gap is less yawning.

Given the tensions separating North and South, and the lesser differences between countries in the North, it would appear that a single set of standards is unlikely to be successful. The Subsidies Code adopted during the Tokyo Round is at least a starting point, but some mechanism must be found to accommodate differences in national priorities linked to levels of economic development and cultural factors.

How might such standards be developed? Consider a 1989 case heard by a panel convened under the U.S./Canada Free Trade Agreement (McRae et al., 1989). The case involved Canadian restrictions on exports of Pacific Coast unprocessed salmon and herring. Although these restrictions date to 1908, they were found GATT-illegal in 1987 after the United States complained that they were unjustifiable restrictions on trade. In 1988, Canada accepted the GATT finding, but stated that it would continue a landing requirement for foreign boats that would allow inspection of their catch. The ostensible reason for this was environmental: it would allow the fish harvest to be counted and monitored so as to preserve the fishery from overexploitation.

According to the U.S., the requirement constituted an export restriction, not only because of the extra time and expense U.S. buyers must incur in landing and unloading, but also because of the dockage fees and product deterioration. The Canadians held that they were pursuing "conservation and management goals" for five varieties of salmon (some of which had previously not been covered by the landing requirement) as well as herring. They justified their action under Article XX of the GATT (the "General Exceptions" section noted above) by appealing to an environmental claim under Article XX(g): conservation of exhaustible natural resources.

The U.S. argued that although the new herring and salmon regulations "are carefully worded to avoid the appearance of creating direct export prohibitions or restrictions, their clear effect is to restrict exports" (McRae et al., p. 13). It was further argued that the Canadian landing requirement was not "primarily aimed" at the conservation of herring and salmon stocks, which had been the interpretation given to Article XX(g) by the 1987 GATT ruling. Thus, the U.S. held that the landing requirement was in fact a restriction on international trade masquerading as an environmental policy, while Canada continued to claim that it was "primarily aimed" at the conservation of the salmon and herring fisheries.

In a significant decision, the panel found that if the effect of such a measure is to impose "a materially greater commercial burden on exports than on domestic sales," it amounts to a restriction on trade, whether or not its trade effects can be quantitatively demonstrated. The panel "was satisfied that the cost of complying with the landing requirement would be more than an insignificant expense for those buyers who would have otherwise shipped directly from the fishing ground to a landing site in the United States" (McRae et al., p. 25). With regard to the Article XX(g) exception, the panel was conscious "of the need to allow governments appropriate latitude in implementing their conservation policies," and also of the fact that the trade interests of one state should not be allowed to override the "legitimate environmental concerns of another" (p. 29). "If the measure would have been adopted for conservation reasons alone," the panel found, "Article XX(g) permits a government the freedom to employ it." Balancing this, however, is the "primarily aimed at" test, which determines whether the measure is part of a genuine conservation or environmental policy, or is in fact a disguised barrier to trade.

This line of reasoning led the panel to two conclusions. First, "since governments to not adopt conservation measures unless the benefits to conservation are worth the costs," the magnitude of costs to the parties—foreign and domestic—who actually bear them must be examined. Second, "how genuine the conservation purpose of a measure is, must be determined by whether the government would have been prepared to adopt that measure *if its own nationals had to bear the actual costs of the measure*" (McRae, p. 31, emphasis added). In this case, the panel was unconvinced that the measure would have been imposed on all Canadian boats primarily for conservation reasons. Specifically, the panel found that Canada would not have adopted such a measure "if it had required an equivalent number of Canadian buyers to land and unload elsewhere than at their intended destination" (p. 32). Alternative methods of monitoring catch rates were available, which posed far fewer restrictions on trade.

Generalizing from this case, it seems possible to envision the development of criteria based on (a) estimated costs of health, safety, and environmental regulations; (b) evidence on who bears these costs; and (c) judgments of whether such measures would be imposed in the absence of any trade effects. Such criteria can serve as a basis for the development of standards to determine which environmental and health measures constitute unnecessary obstacles to trade.

In view of differences in levels of economic development and national priorities, it is clear that these standards cannot be wholly uniform. Jeffrey James, in *The Economics of New Technology in Developing Countries* (1982), suggests that despite valid arguments for improved health and environmental regulations in the South, "it does not follow from this that countries of the Third World should adopt either the same *number* or the same *level* of standards as developed countries." James suggests what may be called *intermediate* standards," in the same sense and for the same basic reason as that which underlies the widespread advocacy of intermediate technology in the Third World." This does not imply a "downgrading" of U.S. regulations, but an "upgrading" of LDC norms, together with recognition that the social costs of regulation are relative to national income.

Under GATT law, these distinctions are recognized as "Special and Differential Treatment" (S & D) of lower income countries. While S & D often creates serious long-run distortions, the terms under which it is granted, as James emphasizes, may actually reduce current regulatory differentials by raising norms in the South, thus improving Third World environmental policies. Although this may not satisfy all competing producers in the North, it can contribute to reductions in overall trade tension, while improving environmental quality in the South.

Conclusion

This chapter has reviewed three somewhat disparate elements of the Uruguay Round negotiations in agriculture. It has argued that a framework agreement on general principles will be reached, against a background of agricultural policies in the U.S. in which Farm Bill decisions are made primarily for domestic reasons. Domestic U.S. interests are not likely to give GATT much leeway, and are threatened by many elements of the negotiation, notably "tariffication" and "rebalancing." Despite these fears, an eventual compromise is expected that will move all parties modestly in the direction of trade reform. There are growing threats, however, to more liberal international trade, arising from new sources that will demand attention. These include rising concerns over health, safety, and environmental quality, which can be turned easily into restrictions on market access.

References

Abel, Daft, and Early. "The Case for Planting Flexibility: An Oilseed Perspective." Washington, D.C.: Oilseed Council of America, January 1990.

Congressional Budget Office. *Farm Program Flexibility: An Analysis of the Triple Base Option.* Washington, D.C., December 1989.

Coughlin, C.C., and K.C. Carraro. "The Dubious Success of Subsidies for Wheat." *St. Louis Federal Reserve Bank Review* 70: 6 (November-December 1988): 38-47.

Hertel, Thomas W. "Negotiating Reductions in Agricultural Support: Implications of Technology and Factor Mobility." *American Journal of Agricultural Economics* (August 1989b): 559–73.

———. "PSEs and the Mix of Measures to Support Farm Incomes." *The World Economy* 12: 1 (March 1989a): 17–27.

Jackson, John H. *World Trade and the Law of GATT.* New York: Bobbs-Merrill Co., 1969.

James, Jeffrey. "Product Standards in Developing Countries." In *The Economics of New Technology in Developing Countries.* Frances Stewart and Jeffrey James, eds. Boulder, Co.: Westview Press, 1982.

Litan, Robert E., and Peter O. Suchman. "U.S. Trade Policy at a Crossroad." *Science* 247 (5 January 1990): 33–38.

McRae, D.M. et al. (Panel Members). *Canada's Landing Requirements for Pacific Coast Salmon and Herring.* Final Report of the Panel convened under the U.S./Canada Free Trade Agreement, 16 October 1989.

Runge, C.F. "Agricultural Trade in the Uruguay Round: Into Final Battle." In *Increasing Understanding of Public Problems and Policies—1989.* Oak Brook, Illinois: Farm Foundation, 1990.

———. "The Assault on Agricultural Protectionism." *Foreign Affairs* (Fall 1988): 133–50.

———. "Induced Agricultural Innovation and Environmental Quality: The Case of Groundwater Regulation." *Land Economics* (1987): 249–58.

Runge, C.F., R.D. Munson, E. Lotterman, and J. Creason. *Agricultural Competitiveness, Farm Fertilizer and Chemical Use, and Environmental Quality: A Descriptive Analysis.* University of Minnesota: Center for International Food and Agricultural Policy, January 1990.

Runge, C.F., and Richard N. Nolan. "Trade in Disservices: Environmental Regulation and Agricultural Trade." *Food Policy* (February 1990): 3–7.

Tangermann, Stefan. "Options and Prospects: A Feasible Package." Paper presented at a conference on "Agriculture in the Uruguay Round of GATT Negotiations: The Final Stages," University of Guelph, Ontario, Canada, 20 February 1990.

8

A New World Agricultural Order?

Murray Fulton and Gary G. Storey

Introduction

The authors of the earlier chapters in this book have provided an understanding of the forces that have shaped the agricultural policies of Europe and North America (Tracy, Skogstad, Rausser), as well as illustrating the effects of these policies on agriculture (Brinkman, Veeman and Veeman). Josling and Runge have focused attention on the most important current issue, namely GATT, which represents an attempt to come to terms with the consequences of the agricultural and trade policies that have been in place over the years, and an effort to establish a "new order" for agriculture.

The overall purpose of this chapter is to examine the dynamics of agricultural policy over the last 200 years—and more specifically the last 25 years—as a basis for understanding the likelihood and the repercussions of alterations to the current order governing agriculture. To undertake this task, the historical transition of agricultural policy and trade in North America and Western Europe is traced, and the economic and political conditions that govern the relationships between these two regions are modelled.

To accomplish this task, the chapter is organized into three distinct sections. The first traces the evolution of agriculture policy from approximately 1800 to the post–World War II period. Its purpose is to illustrate that agriculture has been characterized by numerous "orders" over the years, and that an understanding of how these orders have evolved requires knowledge of political and economic events.

The second section leaves the broad historical arena and focuses on the relationship between the United States (U.S.) and Western Europe during the period from the early 1960s to the late 1980s. In this section it is argued that events during the last twenty five years can be modelled and explained using a political economy framework. Using this framework, the difficulties of making major changes to the current order are examined.

The third section concludes the chapter by drawing together the lessons learned from the previous two sections. While agricultural policy has evolved over the years, and will continue to do so, at present many of the potential orders for agricultural policy have to be considered unreachable. Although a new order in U.S. and European Community (EC) agricultural trade is possible, a number of changes will have to occur.

The Evolution of Agricultural Policy

The transformation of agriculture in Western Europe and North America over the last 200 years has been tremendous. Among the many changes that have occurred are an increase in productivity, the opening up of new agricultural lands, and a major migration of labor out of the industry. One of the underpinnings of this transformation was the ascendency of the philosophical notions of liberalism, individual rights, and a market economy, which had their roots in the writings of people like Locke and Adam Smith.

The most important issue for this chapter, however, is not the developments per se, but rather the manner in which society has dealt with these changes economically and politically. In particular, the underlying themes of this chapter are an examination and a recognition of the distributional questions regarding who in society will bear the cost and receive the benefits of this transformation.

Framed in this manner, the evolution of agricultural policy can be seen as a result of different groups in society attempting to gain or retain power in an effort to influence decisions regarding their welfare. In the context of this chapter, the notion of an economic and political "order" is to be interpreted as a particular distribution of income and wealth that arises from the agricultural policy decisions that have been made.

In its narrowest sense, an order involves the distribution among groups within a country or region, e.g., consumers, farmers, taxpayers. More broadly, however, an order involves the distribution among different countries or regions of the world. Since the economies of different countries are linked through international trade, attempts by one country to obtain additional benefits for its citizens will have an effect on the citizens of other countries.

Although this chapter deals with the intercountry aspect in the context of North American and Western European trade, it should be pointed out that the distributional questions concerning countries outside this region (e.g., developing countries, the food importing countries, the food exporting countries) have largely not been examined.

As Polayni points out in *The Great Transformation* and *The Livelihood of Man*, governments generally have not allowed the free market to make the decision as to who should bear the cost of economic progress. As will be seen, the same point can be made with respect to agriculture. Indeed, one of the conclusions that will be drawn from the historical overview is that agricultural protection has generally been the rule, rather than the exception.

The world has tried a diverse set of orders in an attempt to deal with political and economic factors. This has also been true for the democracies of Western Europe and North America, which are the focus of this book. While some have been more successful than others, none has been stable or lasting. Indeed, another theme that emerges in this section is that agricultural policy is inherently unstable. As discussed in the conclusion, this presents a particular challenge for policy makers in the future.

Basic to the discussion in this section is the notion that major shocks—economic depressions, wars, the introduction of new technology—have resulted in new policies toward agriculture and trade. These shocks are illustrated in Figure 8.1, along with the major changes that have occurred in agricultural policy in North America and Western Europe. Figure 8.1 thus serves to summarize the material to follow in the remainder of this section.

The Period 1800 to 1850

One of the major constraints to industrial growth in Europe at the end of the eighteenth century was the protection of agriculture. In Great Britain, the Corn Laws provided tariff protection for grains. This meant high prices for grain, and thus high prices for bread, which was the staple food. Given the subsistence level of the working classes, the wage rate was largely a function of the price of bread. Thus, increased profits for industry required a reduction in the price of grain, which meant reducing agricultural protection. As argued by political economists and parliamentarians like David Ricardo, it was inconsistent to apply principles of free trade only to manufacturing. He argued, as had Smith, that the mercantilist policy of supporting exports and restricting imports in order to amass money aswealth was false economics. The basis of wealth was seen as increased production stemming from specialization, which by the principle of comparative advantage was extended internationally through trade.

Figure 8.1 Chronology of European-North American Agricultural Policies, 1800-1990

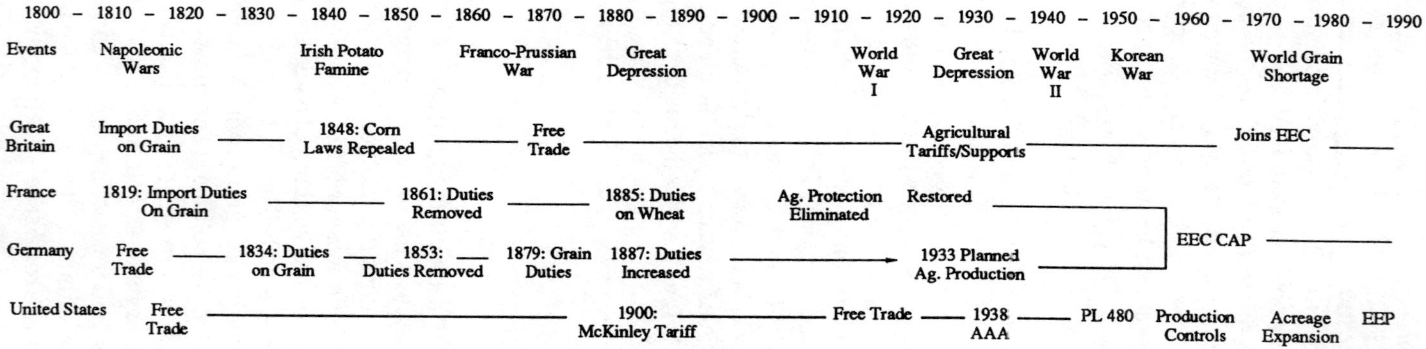

	1800	1810	1820	1830	1840	1850	1860	1870	1880	1890	1900	1910	1920	1930	1940	1950	1960	1970	1980	1990
Events	Napoleonic Wars				Irish Potato Famine			Franco-Prussian War	Great Depression			World War I		Great Depression	World War II	Korean War			World Grain Shortage	
Great Britain	Import Duties on Grain				1848: Corn Laws Repealed			Free Trade												
France			1819: Import Duties On Grain				1861: Duties Removed		1885: Duties on Wheat		Ag. Protection Eliminated	Agricultural Tariffs/Supports Restored								
Germany	Free Trade			1834: Duties on Grain		1853: Duties Removed			1879: Grain Duties	1887: Duties Increased				1933 Planned Ag. Production			EBC CAP	Joins EEC		
United States	Free Trade										1900: McKinley Tariff		Free Trade	1938 AAA	PL 480		Production Controls		Acreage Expansion	EEP

The extension of voting privileges in 1832 to a large segment of the commercial class, and the potato famine in 1845, led to the repeal of the Corn Laws in 1848. As a result of reduced protection, Great Britain's farmers experienced a decline in grain prices, grain production, and land values, and the country became a major food importer. Great Britain had decided to sacrifice the interests of landowners and farmers in order to further the growth of the economy through expanded industrial output and trade.

In France and Germany, agriculture received little protection at the turn of the century, but this changed in 1819 when France introduced a sliding scale of duties on imported grain similar to the British Corn Laws. Protection was reduced following the Napoleonic wars in Germany, but tariffs on grain imports, although moderate, became part of the Zollverein tariff introduced in 1834. In Canada and the U.S., agriculture was left unprotected.

The Period 1850 to 1900

In Europe the period 1850 to 1900 was characterized by a struggle between industrial and agricultural interests (landowners primarily) over free trade. The struggle became particularly critical with the economic depression after 1870. The decline in grain prices in Europe due to increased grain imports from the U.S. resulted in a return to protection in Germany and France, where the farm interests had more power. Great Britain, however, maintained its position on free trade and laissez-faire. The following are some of the major events and developments.

The repeal of the Corn Laws in Great Britain in 1848 launched agriculture into an era of free trade that was to last until the 1930s. It also set in motion changes in continental Europe. Under the leadership of France, various treaties were signed encompassing the "most favored nation" principle, which resulted in agricultural products having lower tariffs than manufactured goods. France abolished duties on grain in 1861, a step Germany had taken in 1853. Similar developments also occurred in Italy, Belgium, and the Netherlands.

Although agriculture in Great Britain prospered immediately following the repeal of the Corn Laws—mainly through improvements in productivity—it was the increase in imports of cheap grain from North America (and Russia) after 1870 that led to a decline in grain prices. This, coupled with bad harvests in Europe, led to calls for renewed agricultural protection during the period 1870 to 1895. By 1913, tariffs on foodstuffs ranged from 20 to 30 percent. With the McKinley Tariff of 1890, even the U.S. extended protection to agriculture. This, however, had no appreciable effect on prices or agriculture since the U.S. continued to be a net exporter.

Although farmers were being ruined in Great Britain, the Netherlands, and Denmark, governments refused to restore protection to agriculture. This forced structural changes. In Denmark and the Netherlands, farmers switched from grain to livestock. Elsewhere in Europe, the renewed tariffs mainly benefited the grain farmers, who tended to be the large landowners, rather than the small livestock farmers.

Tracy (1984) indicates that over the period of the economic depression (1880-1900) crop prices in Great Britain averaged 68 percent of the predepression level (1867-77). This fall in prices gave rise to several commissions of investigation, which failed, however, to recommend that protection be restored. Great Britain remained a committed free trader until the 1930s, even in the face of renewed protection on the continent. The policy took its toll, however, as the agricultural population fell from 18 percent of total population in 1861 to 8 percent in 1901. Farm income in the 1892-96 period was only 72 percent of what it was in the period 1867-78.

In contrast, the increase in protection for French agriculture after 1880 had the effect of raising prices, increasing production, and making France a net exporter of food. In Germany, agriculture prospered in the 1850-1870 period, with the country being a net exporter of grain. Although farmers were initially opposed to any protection, industry was not. Following the decline in grain prices that occurred when U.S. exports to Europe increased, the farming interests in Germany sided with industry and demanded protection, which began in 1879, with further tariff increases in 1885 and 1887. Following the expansion of industry in Germany, and the subsequent need for export markets, industrialists called for free trade. This paralleled developments in Great Britain prior to the repeal of the Corn Laws.

Tracy explains that with Imperial Germany interested in expansion, its policy of rearmament went hand-in-hand with increased economic self-sufficiency in all sectors, including agriculture. Of particular interest in this period was the concern that the country was losing its agricultural population to industry, with a corresponding greater dependence on the export market for industrial goods and an increasing reliance on the importation of foodstuffs. Food security became a major issue. Protection was seen as a basic long-term need to maintain a balance between industry and agriculture. These Free Traders, as explained by Tracy, argued "that only a minority of the agricultural population—the large landowners—

benefited from high grain prices" (Tracy, 1984). The balance of political power, however, was on the side of the protectionist forces.

The Period 1900 to the 1930s

This period is important for the changes that took place in the structure of the international grain economy. The increase in grain production in the New World that had been hoped for by the industrial groups after the repeal of the Corn Laws, and the move to free trade, finally came about in the early part of the twentieth century. The cheap food that the industrialists wanted occurred as a result of low grain prices, but with such devastating consequences for farmers in the 1930s that it gave rise to a new political and economic philosophy toward agriculture in both Europe and North America. The following are some of the main highlights of this period.

For Great Britain to reduce the price of food, it was essential that imports be increased. This was to be accomplished through the importation of grains from the New World, which meant that the agricultural frontier had to be expanded westward in North America and new lands be developed in Argentina and Australia. Palliser and Hind were sent by the British and Canadian governments respectively to investigate the feasibility for agriculture and settlement in the western territories of North America. They reported in the early 1860s. Hind concluded that a lack of markets was the major impediment to settlement, but in fact, the lack of transportation and the cost of ocean freight were the real limiting factors. A number of new technologies, however, eventually aided the development of the wheat economies in western Canada and elsewhere.[1]

The growth of the grain economies of the U.S., Canada, Australia, and Argentina is best illustrated by the expansion in seeded acreage and production between 1885 and the 1930s (see Table 8.1). The data shows that only the U.S. was a significant grain producer at the end of the nineteenth century. By the 1930s, however, seeded acreage in all four countries had increased from 43.7 million acres to 125.1 million acres.

The First World War, which disrupted agricultural production in Europe, was instrumental in raising world agricultural prices. While tariffs on agricultural products were suspended during the war, they were reintroduced once it was over. Production from the opening of new grain lands in North America, Australia, and Argentina continued to increase during the 1920s. By 1929, world grain stocks reached burdensome levels. Similarly, industrial production had outstripped demand, resulting in the

Table 8.1 Seeded Acreage of Wheat—United States, Australia, Canada, and Argentina

	United States	Canada	Australia	Argentina	Total
			millions acres		
1885-89	35.81*	2.58	3.31	2.00	43.70
1899-1904	49.42*	4.14	5.38	8.87	67.81
1922-27	61.97	22.06	10.40	17.97	112.40
1932-37	68.15	25.47	13.50	17.99	125.11

Source: de Hevesy
* Harvested

collapse of stock markets in October 1929. This set in motion a wave of tariff escalation as countries attempted to isolate their economies.

Following the economic collapse, exporting countries attempted to dump surpluses with the aid of subsidies. Importing countries in Europe countered with increased tariffs. In Germany and France, the internal price became two to three times the world price of grain. When tariffs were ineffective, countries adopted nontariff means of restricting imports. France, Germany, and other European countries, for example, required millers to use a minimum percentage of domestically produced wheat.

The severity of the Great Depression of the 1930s was instrumental in fostering a major change in the economic and political philosophy toward agriculture. Great Britain abandoned its policy of free trade and resorted to protection. There was the realization that agriculture should not be subjected to the rule of the free market. De Hevesy (1940) stated, "the collapse of the world wheat price on the one hand, and the social, economic and political importance of the farmer and peasant on the other, are the reasons which have, for the past few years, obliged every government in the world to provide some sort of financial aid to its wheat growers." He also observed, however, that "Subsidized farming fosters excess production, excess production necessitates subsidized exports, and subsidized exports always exert a depressing effect on the world price, which in its form is detrimental to the farming 'interest.' " This was to be a truism of the latter postwar period.

A New World Agricultural Order?

The Period 1930s to 1990

This period manifests the results of the new philosophy adopted by governments in both North America and Europe to extend support to farmers and the agricultural sector. With the introduction of the Common Agricultural Policy (CAP), Europe abandoned its cheap food policy. Although the policy was not given up in North America, governments did relinquish their laissez-faire attitude toward agriculture and took an interventionist position to support farm prices and incomes. This section presents the major events that complement the next main section of the chapter.

The agricultural crisis of the 1930s was recognized as one of overabundance in the face of a general inelastic demand for food. In the U.S., Secretary of Agriculture Wallace attempted to reverse the protectionist agricultural policies that had developed, but he did not call for a complete return to free trade with no government involvement. As Hadwigger (1970) explained, "Wallace envisioned a world economy, but not of the traditional laissez-faire order which he felt could lead only to further cycles of depression and war. His world economy would be a long-run product of education and laborious planning." It was a mistrust of the free market that led Wallace to propose this essentially new doctrine for a planned world agricultural economy, the idea of the "ever normal granary" that would serve to buffer and moderate the price cycle.

Although Wallace's ideas did not find general acceptance internationally, they had two effects. One result was the convening of the London Conference, with the signing of the International Wheat Agreement of 1933. This was an attempt to adjust supply to effective demand and to raise and stabilize prices to a level remunerative to farmers and fair to consumers. The agreement, although failing in the 1930s, was the forerunner of the International Wheat Agreements that were in effect from 1949 to 1967.

The second result was the formation and passage of the U.S. Agricultural Adjustment Act of 1938. It contained provisions for a price support loan program based on parity price, programs to expand agricultural usage and exports with subsidies if required, and the use of production controls to prevent stock accumulations. The act contained within it all the seeds of U.S. agricultural policies from the 1950s to the present day. Although the specifics have changed, the overall policy remains in place.

After the Second World War, the U.S. had a dominant influence in framing the new institutions, including the United Nations, Bretton Woods, and the GATT. It used its power to exclude agriculture from the GATT.

The implications this has had for agriculture and trade policy has been discussed by Josling in this volume.

In the 1950s, the U.S. was to experience first-hand the problems of surplus production emanating from price supports set higher than world market levels. As a result, the U.S. was forced to introduce export subsidies and production controls as a means of dealing with the surpluses. However, acreage controls encouraged more intensive farming and thus resulted in increased yields, largely through improved technology, which further contributed to the surplus problem. What is significant about the 1952 to 1960 period is that the reduction in support prices (although modest) and acreage controls failed to reduce production. Even food aid exports under Public Law (PL) 480 failed to solve the problem.

The high price structure established under parity during the Second World War proved hard to demolish. Farm interests were simply too powerful and entrenched in Congress. The 1960s can be characterized as an attempt by a new administration (now Democratic) to gain greater control over production and thus surpluses, while at the same time continuing to support farm incomes. This, however, did not reduce the trend toward increased government involvement. The government felt responsible for the problem because of its investment in the development and introduction of new agricultural technology, which resulted in increased productivity.

The major objectives of the U.S. administration in 1960 were to raise farm income, reduce government costs, use surplus food constructively for agricultural development and promote the efficient use of land resources. The administration was able to introduce more rigid production controls, which resulted in a reduction in wheat acreage in the mid-1960s. One of the more significant departures from former policy was the introduction of the domestic wheat certificates in the Food and Agricultural Act of 1962, which was a tax on domestic processors that was passed on to consumers as higher bread prices. In essence, it shifted some of the burden of supporting farm incomes from taxpayers to consumers. This was similar to the approach adopted by the EC for its new agriculture policy. The certificate program was removed in 1967.

Before discussing the development of the CAP, it is important to assess the policies that were in place and the philosophy that prevailed in the Western European agricultural sector. The 1957 Treaty of Rome provided the basis for a common market, thus removing trade barriers between member states and establishing a common external tariff. Agriculture posed a problem. The treaty provided for the establishment of a common agricultural policy, but there were wide differences among the six members in terms of the structure and the productive efficiency of agriculture in each country. The critical issue of agricultural prices was resolved in favor of Germany, which had the highest food consumption and price levels.

Although agricultural surpluses were a problem, the emerging policy did not effectively come to grips with it. There was more interest in assisting developing countries with surplus food than in imposing production controls. A philosophy of market intervention and protection prevailed. With goals of increasing farm incomes, encouraging adjustment and increasing productivity, it should not be surprising that production expanded and created surpluses. Consideration of "fair prices" for food for consumers took second place to the goals associated with supporting farm income.

For the purposes of the analysis it is not essential to outline the details of the CAP, which are available from a number of sources, including Fennell and Harris, et al. The entry of the United Kingdom, Ireland, and Denmark in 1973 and more recently Greece, Portugal, and Spain to the EC, has had little direct bearing on the existing agricultural policy, as new members have tended to adopt the CAP programs already in place.

The EC reached the turning point in 1982/83 when it became a net exporter of grain. In 1977/78 it had net imports of 18 million tonnes; in 1988/89 it had net exports of approximately 25 million tonnes. As a result, export subsidies increased from 1,076 million ecus in 1985 to 3,518 million ecus in 1988.

Without having solved its farm problem, agriculture was absorbing over 60 percent of the Community's budget. The EC found itself having to compete for markets for its surplus cereal and other products, against countries that had a more legitimate claim to being exporters from principles of comparative advantage. This surplus production thus contributed to lowering world prices and forcing exporting countries like Canada, the U.S., Australia, and New Zealand into developing domestic policies of agricultural adjustment or attempting to protect their farmers through increased subsidies.

A pivotal point in U.S./EC relations over agriculture occurred in 1982/83 when the EC shifted to becoming a net exporter of grains. What did the EC expect the U.S. to do in response to its attack on the U.S. and other exporters' markets? It may have expected the U.S. to cut back its own production to accommodate declining export markets. This may have been a reasonable expectation, as the U.S. had tended to follow this policy in the early 1980s when it began to lose export market share. However, the U.S. decided to fight for market share and reintroduced export subsidies through what is referred to today as the Export Enhancement Program, first introduced in 1984.

The period from 1985 to 1990 can be depicted as one of escalating trade warfare between the U.S. and the EC. The 1988 drought in North America was primarily responsible for a decline in world grain stocks, and prices increased somewhat. This did not result in a cessation of export subsidization. It seemed obvious that the U.S. wanted to maintain its pressure on the EC in its attempts to obtain a more favorable result for its GATT position eliminating agricultural subsidies.

Summation

The material in this section has focused on the transformation of agricultural policies in Europe and the U.S., the two major agricultural economies. The analysis indicates that agricultural policy has shifted from one of free trade and laissez-faire in the middle of the nineteenth century to one of extensive protection and government involvement in the latter part of the twentieth century. This can be interpreted as a search for the right "order" for agriculture. The transformation can be illustrated in a two-region model (see Figure 8.2), which is also designed to indicate possible future orders for agriculture.

Figure 8.2 Two-Region Policy Transition Model, 1850-1990

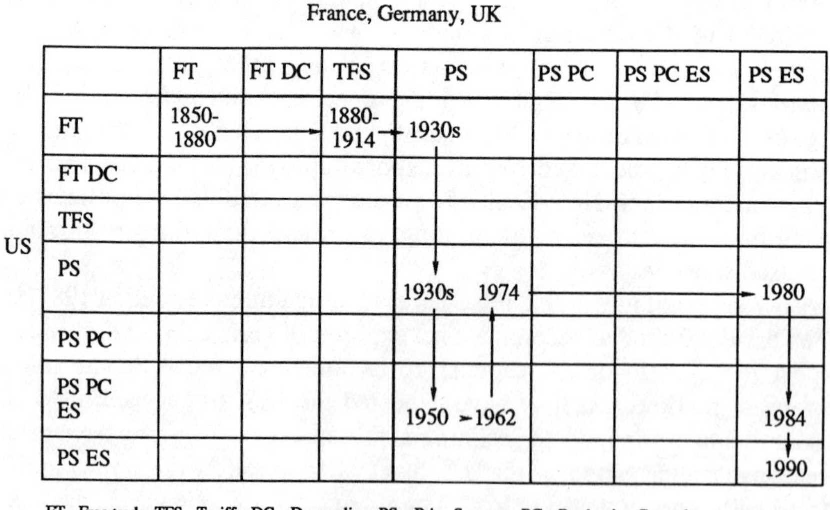

France, Germany, UK

		FT	FT DC	TFS	PS	PS PC	PS PC ES	PS ES
	FT	1850-1880		1880-1914	1930s			
	FT DC							
	TFS							
US	PS				1930s 1974			1980
	PS PC							
	PS PC ES				1950 - 1962			1984
	PS ES							1990

FT - Free trade; TFS - Tariffs; DC – Decoupling; PS – Price Supports; PC – Production Controls; ES – Export Subsidies

The model has been simplified to include six major policy options: free trade (FT), tariffs (TFS), decoupled programs (DC), price supports (PS), production controls (PC), and export subsidies (ES). The model begins in the upper left corner with a situation of free trade in both the U.S. and Europe (France, Germany, and the United Kingdom). This model indicates that Europe introduced tariffs to protect its farmers in the period 1880 to 1914. As explained previously, this development was primarily based on the continent, with Great Britain (UK) maintaining a free trade policy. As described, laissez-faire was abandoned in the 1930s in favor of government intervention in both Europe and the U.S. This is depicted as the introduction of price supports. With high price supports, the U.S. quickly developed grain surpluses and had to resort to export subsidies (PL 480) and production controls. This came in the 1950s. With the introduction of the CAP in 1962, the EC put in place a program that would include production controls and export subsidies. Price supports were raised significantly from previous levels for most of the new members of the EC. This is seen as a further escalation of government intervention. With the tightening of world stocks and the high prices that resulted in 1973 and 1974, the U.S. cancelled its export subsidies and reduced production controls.

The next major transition occurred between 1974 and 1980 when the EC expanded its export subsidies to deal with its growing grain surpluses. The U.S. initially accommodated this move and let the EC capture an increasing market share of a declining export market. In 1984, however, the U.S. broke with this policy and reintroduced export subsidies. The trade war was on, thus further escalating government intervention. The U.S., however, continued to temper its program with production controls, but with low U.S. stocks, production controls were essentially removed for the 1990 crop year. Even though the EC has placed some limitations on production, it is still questionable how effective they will be.

The model thus illustrates a diagonal transition from the upper left to the lower right in terms of the degree of government intervention in agricultural affairs. This can be seen as a search for the right order for agriculture, given established goals for farmers, consumers, and government itself.

The model has relevance for future agricultural orders. The Uruguay Round is seen as a negotiated search for a new agricultural order. The solution would see a return to free trade—the 1850 situation. Others see free trade in conjunction with continued support for farmers through the direct payments of decoupled programs. A failure of GATT negotiations would see a possible continuation of the trade war, until one or both the U.S. and/or the EC adopted policies to better constrain production and subsidies unilaterally. These possibilities are explained in the next section of the chapter, where a more rigorous model is developed to explain the agricultural policy transition of the past twenty-five years.

A Political Economy Model of U.S./EC Policy and Trade

The purpose of this section is to examine the agricultural policies of the EC and the U.S. from a political economy perspective. Given space constraints, the focus of the discussion will be on one particular commodity—wheat. While the conclusions derived from the analysis in this section will not necessarily be applicable to other commodities, the methodology presented here can be used to examine the other elements of U.S./EC trade.

It will be argued in this section that it is possible to explain the policies that have been introduced in the U.S. and the EC over the past twenty-five years by examining internal political factors in these regions, and by taking into consideration the fact that both regions are large enough that their internal policies will have an impact on the world market. In short, the economic order of the last three decades is explainable in economic and political terms.

To the extent that the model developed is able to explain the historical pattern of domestic agricultural policy in these two regions, it can then be used to examine the consequences of alterations to the current order. The purpose here is to analyze the extent to which the two regions will be willing to alter their agricultural programs and to agree to a new international order. The conclusion reached is that while there is likely some room for an agreement, the changes associated with it are more a matter of degree than they are of kind.

Model Assumptions

The model developed in this section is based on a number of assumptions. The first is that the objective of governments in the U.S. and the EC is to maximize the returns to farmers, less the cost to government. Support for this notion is given by Michael Tracy and Gordon Rausser, who argue that producers in both regions have, for a variety of reasons, been able to capture the government. Adopting this assumption explicitly recognizes that consumers in the U.S. and the EC have little political power.

The second assumption is that agricultural policy in the U.S. is designed so that consumers in that region pay the world price, while the agricultural policy of the EC is structured so that consumers pay the internally set price. This assumption reflects not only what has historically been the case,[2] but also the different political economy in the two regions. More specifically, the assumption that EC consumers pay the internally set price indicates that consumers in this region have been willing to accept such a policy, while their U.S. counterparts have not. Among the reasons for this willingness are a desire for food security, a lack of political power, and the recognition that even with such a policy in place, food costs represent a declining share of total household expenditures. Thus, although consumers in both regions have little political power, those in the U.S. have generally been able to pay the lower world price.

As will be seen, the difference in consumer prices in the two regions represents an asymmetry that plays a fundamental role in explaining the policies of the two regions over time. In addition, the assumption that EC consumers are willing to pay the internal price means that decoupled programs are effectively ruled out. More specifically, as long as consumers are willing to finance agricultural programs through the price they pay for food, there is no incentive for the EC to alter this and obtain the money required for government programs from taxpayers.

A third assumption is that U.S. and EC exports affect the world price. There are two components to this assumption. The first is a recognition that the internal policies of the major trading regions have a spill-over effect on the world market—i.e., that the external effects of domestic policies are responsible for the volatility and level of prices in the export market (Carter, McCalla, and Schmitz). The second involves a recognition by the U.S. and the EC that their internal policies will affect the world price. More specifically, if the world price is affected by internal policies, and the level of the world price influences the ability of a region to achieve its domestic goals, then it is assumed that both the EC and the U.S. will adjust their domestic policies accordingly.

The final assumption is that growth in agricultural productivity is greater in the EC than in the U.S. This again reflects the historical record, where yields in the EC have increased by 3.6 percent per year over the period 1960-85, compared to a 1.4 percent increase in the U.S. (Carter, McCalla, and Schmitz).

Historical Review

The purpose of this section is to examine the events that have occurred over the last twenty-five years in the world wheat market and to show that they can largely be explained with the use of the model presented above.

Figure 8.3 illustrates the state of the world wheat market when the CAP was formed in the early 1960s. The net demand for wheat by all countries outside the U.S. and EC—demand curve D^1 in the export market—is presented in the third panel of Figure 8.3. Given knowledge of the imports and exports of the two regions, the world price, p_w^1, is determined from this demand curve.

The imports and exports of the U.S. and the EC are a function of the internal policies in these regions. In the EC, the domestic demand curve (D_{ec}) is assumed to be price inelastic,[3] while the domestic supply curve (S_{ec}^1) is a function of the level of domestic price, p_{ec}. The establishment of an internal price that is above the world price leads to a reduction in the level of imports from what they would have been in the absence of such a program.[4] As illustrated in the second panel of Figure 8.3, imports into the EC (m_{ec}^1) are determined by the difference between domestic demand (y_{ec}) and domestic supply (z_{ec}^1).

In the U.S., the domestic demand curve (D_{us}) is also assumed to be price inelastic. Although a domestic supply curve (S_{us}) exists, production (and

Figure 8.3   The U.S. and EC Wheat Market in the 1960s

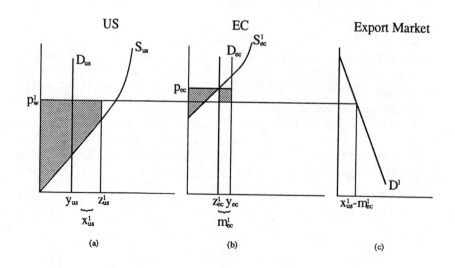

hence exports) in the U.S. is not always determined by either the world price or the domestic price. This is because an important part of U.S. policy over the past thirty years has been acreage set-aside programs and government stock-holding. Indeed, Figure 8.3 indicates that in the 1960s the U.S. was not operating on its supply curve, but had reduced its production (z_{us}^1) through acreage reduction, which meant that exports (x_{us}^1) were smaller than they would otherwise have been.

U.S. and EC policies reflect attempts by these two regions to pursue an objective of maximizing producer returns, less government payments. As McCalla points out, it is advantageous for the U.S. to restrict its output to some extent, thereby raising the world price above what it otherwise would be.[5] At the same time, it was desirable for the EC to increase its internal price above the world price, thereby increasing production. This particular question will be examined in more detail below.

In the U.S., the level of producer returns (less government expenditures) is given by the shaded area in panel (a) of Figure 8.3. In formal economic terms, this area represents the producer surplus accruing to U.S. farmers. For the EC, the level of producer returns (less government expenditures) is given by the shaded areas in panel (b). The triangular area below p_{ec} and above S_{ec}^1 represents the returns to producers, while the rectangular area ($m_{ec}^1(p_{ec} - p_w^1)$) represents the revenue received by the EC government from importing wheat at the world price and selling it domestically at the internal price.

Figure 8.4 models the world wheat economy in the mid-1970s and presents a comparison with the situation in the 1960s. A distinguishing feature of the mid-1970s was the large outward shift of the export demand curve (D^1 to D^2) as a result of poor crops and increased demand in areas outside the U.S. and the EC. More specifically, the shift in D represents the impact of greater demand by the developing countries, and reduced production in countries like China and the USSR. Compared to the 1960s, the 1970s also saw an outward shift of the supply curve in the EC and the U.S., a result of increased agricultural productivity. As noted above, this shift was greater in the EC than in the U.S., a differential modelled in Figure 8.4 by an outward shift of the EC's supply curve (S_{ec}^1 to S_{ec}^2), while the U.S. supply curve remains stationary.[6]

With the outward shift of the supply curve, output in the EC increased from z_{ec}^1 to z_{ec}^2, and the EC became essentially self-sufficient (z_{ec}^2 is approximately equal to y_{ec}). Table 8.2 shows the pattern of production and exports (imports) over the period from the mid-1960s to the mid-1970s.

Figure 8.4 The U.S. and EC Wheat Market: Mid-1970s versus the 1960s

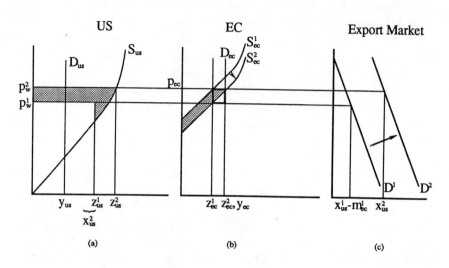

During this period of time, the increase in export demand meant that despite the increase in output, the market for U.S. exports expanded. To meet this demand, the U.S. expanded production and reduced its level of stocks; acreage controls were removed and farmers were encouraged to plant "fence-row to fence-row." The result was that U.S. production was determined largely by market forces. With a world price of p_w^2, U.S. output increased to z_{us}^2, leading to an increase in U.S. exports (x_{us}^2 versus x_{us}^1). Table 8.2 presents data illustrating these changes.

This state of affairs was advantageous to both regions. With a strong export demand, it was no longer in the best interests of the U.S. to restrict output. More precisely, perhaps, while the U.S. found it beneficial to restrict output, the level to which they wished to restrict it was greater than the maximum level of output they were able to produce with a full production policy. The notion that the U.S. was constrained in its production capability was reflected in the literature on capacity constraints in agriculture that emerged during the mid-to-late 1970s.[7]

In the EC, the increase in world price to roughly the level of the internal price meant that domestic policies could be retained with little cost. Although the price they received changed very little, producers were better off as a result of having more produce to sell. For example, the supply shortage and the subsequent increase in the world price in the 1970s highlighted the need for a policy that would provide European consumers with some degree of food security.

Table 8.2 Wheat Production, Exports, and Market Share, U.S. and EC, 1960-1988

	United States			European Community		
Year	Production	Net Exports (Imports)	Market Share	Production	Net Exports (Imports)	Market Share
	(mmt)	(mmt)	(%)	(mmt)	(mmt)	(%)
1960	36.9	17.8	41	33.7	(10.3)	(24)
1961	33.5	19.5	41	32.1	(10.1)	(21)
1962	29.7	17.7	38	41.9	(4.7)	(10)
1963	31.2	23.0	40	35.3	(5.8)	(10)
1964	34.9	19.7	36	40.5	(3.1)	(6)
1965	35.8	23.2	38	42.9	(4.0)	(7)
1966	35.5	21.0	36	37.7	(4.4)	(8)
1967	41.0	20.8	39	44.2	(2.1)	(4)
1968	42.4	14.8	29	44.4	(2.8)	(6)
1969	39.3	16.4	29	42.5	(1.4)	(3)
1970	36.8	20.2	36	41.3	(6.9)	(12)
1971	44.1	16.6	29	48.3	(2.4)	(4)
1972	42.1	30.9	42	48.4	(0.1)	—
1973	46.6	33.1	45	47.7	(0.6)	—
1974	48.5	27.7	40	52.7	1.9	3
1975	57.9	31.9	43	45.1	2.2	3
1976	58.5	25.8	37	46.6	0.5	1
1977	55.7	30.6	41	44.5	(0.6)	(1)
1978	48.3	32.5	39	55.3	3.6	4
1979	58.1	37.4	40	53.2	5.8	6
1980	64.8	41.2	43	61.5	10.3	11
1981	75.8	48.2	45	58.1	10.0	10
1982	75.3	41.1	38	64.7	11.8	11
1983	65.9	38.9	35	63.8	11.8	11
1984	70.6	38.8	33	82.8	15.3	13
1985	66.0	24.9	25	71.8	12.8	15
1986	56.9	27.3		71.9	14.0	
1987	57.3	43.3		71.6	12.3	
1988	49.6	38.1		76.7	16.0	

Blanks indicate data not available; indicates zero
Sources: C. Carter, A. McCalla and A. Schmitz, Table 2-10; USDA, Foreign Agricultural Service, *World Grain Situation and Outlook*, selected years.

The notion that both regions were better off can be examined graphically in a much more formal manner. Using the returns to producers (less the cost of government payments) as the measure of well-being, the welfare of the U.S. can be shown to increase by the shaded area in panel (a), Figure 8.4. In the EC, producer welfare increases by the shaded area in panel (b), while government revenue falls by the amount of the outlined rectangle. The net change in EC welfare is given by subtracting the government loss from the producer gain.

Following the buoyant times of the mid-1970s, the world grain market declined in the late 1970s and early 1980s. This downturn can be reflected graphically in Figure 8.5, where the export demand curve facing the U.S. and the EC has shifted in from D^2 to D^3. The causes of this shift were numerous, including an increase in supply from some of the traditional exporters (e.g., Canada), as well as a reduction in demand by the traditional importers (e.g., USSR, China).

The result was a fall in world price. This decline, however, was mitigated somewhat by the policy reaction of the U.S. With the fall in export demand, the U.S. returned to its previous policy of holding stocks and reducing acreage. The result was that U.S. output fell to z_{us}^3. Note that this output is less than what the country would produce if it were to simply take price p_w^3 as given and produce off of its supply curve. The reduction in output causes a corresponding decrease in exports (x_{us}^3 versus x_{us}^2). The

Figure 8.5 The U.S. and EC Wheat Market: Late 1970s/Early 1980s versus the Mid-1970s

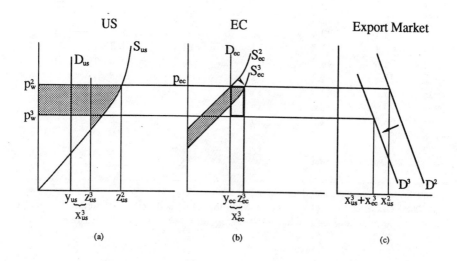

result of these changes was a decrease in the overall welfare of the U.S. by an amount equal to the shaded area in panel (a) of Figure 8.5.

In the EC, the reaction to the shift in export demand led to changes that were the reverse of those in the U.S.. As a result of greater productivity, the supply curve in the EC continued to shift outward (S_{ec}^2 to S_{ec}^3). The EC did not change its internal policy to any great extent, i.e., the internal price remained relatively constant. With the shift in supply, production increased to z_{ec}^3, which resulted in the EC becoming a net exporter of wheat, with exports equal to x_{ec}^3. With the shift of the supply curve, the welfare of producers increased by the shaded area in panel (b) of Figure 8.5.[8] At the same time, however, EC government expenditures increased by the highlighted rectangle (area $x_{ec}^3(p_{ec} - p_w^3)$), as it instigated export subsidies to compete in the world market.

There are a couple of important points to be made with regard to this period. The first is that despite being very different, the internal policy responses of both regions were nevertheless optimal for the situation in which they found themselves. As discussed above, the U.S. found it desirable to reduce its production and increase its level of stockholding in an attempt to keep the world price from falling as low as it might otherwise have done. In other words, the U.S. behaved much like an exporter with market power could be expected to behave—in an effort to use its market power, it reduced output.

The EC reaction was just the opposite, despite the fact that it had, for the first time, become a net exporter. In spite of this fact, however, it was desirable for the EC to continue to expand production by keeping the internal price above the world price.

The reason for this behavior lies in the EC consumer's willingness to pay the internal price as opposed to the world price. To see this more clearly, recall that the objective given both regions is to maximize producer returns, less the cost to government. Since consumers pay the world price in the U.S., any attempt by the U.S. government to pay producers a price above the world price will involve a large monetary outlay, since U.S. output is substantial. As a result, producer returns can be maximized by reducing output, since doing so will reduce exports, raise the world price, and increase the returns to producers. The relatively inelastic demand for wheat makes this strategy even more desirable for the U.S.

In the EC, however, this same logic does not hold. Since consumers are willing to pay the internal price, the government does not have to pay the entire cost of raising the internal price above the world price. In fact, by raising the internal price above the world price, the EC government is able to obtain benefits for producers without having to pay for all them itself. Thus, the optimal response for the EC government is to keep the internal price above the world price, even if this increases European production, reduces the world price, and results in the EC government subsidizing its expanded exports.

This can be seen more formally with the help of Figure 8.6. Suppose that the world price is p_w. In the absence of a domestic policy, EC producers earn returns equal to the area below p_w and to the left of S_{ec}, while government expenditures are zero. Suppose, now, that the EC establishes an internal price equal to p_{ec}. This results in an increase in producer returns equal to the shaded area above p_w, below p_{ec}, and to the left of S_{ec}. In order to achieve this gain in producer returns, however, the government must make an expenditure equal to $x_{ec}(p_{ec} - p'_w)$, i.e., the highlighted rectangle. This expenditure reflects the amount the government must incur to make wheat produced in the EC competitive on the world market at a price p'_w. Note that the world price has been lowered (p_w to p'_w), resulting in the higher domestic price and the subsequent increased production in the EC.

The significant point is that in terms of achieving the objective of maximum producer returns less government expenditures, it may be desirable for the EC to adopt an internal price that is above the world price. This is seen in Figure 8.6, where the shaded area above p_w, below p_{ec}, and to the left of S_{ec} (the gain in producer returns) is greater than the area $x_{ec}(p_{ec} - p'_w)$, the increase in government expenditures. In other words, even when its influence on the world market price is accounted for, it may still be optimal for the EC to support producer prices and subsidize exports.

Figure 8.6 Economics of Protection in the EC

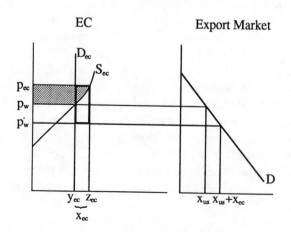

A New World Agricultural Order?

Table 8.3 Historical Wheat Prices, U.S. and EC, 1960-1989

Year	World Price[a]	United States Target Price	United States Loan Rate	European Community Target Price	European Community Threshold Price	European Community Target Price	European Community Threshold Price
	(U.S.$/mt)	(U.S.$/mt)	(U.S.$/mt)	ecu/mt	ecu/mt	£/ton	£/ton
1960	72.80	65.40	63.93				
1961	75.70	65.77	67.24				
1962	82.30	73.49	74.96				
1963	79.00	73.49	69.08				
1964	59.20	73.49	47.77				
1965	58.80	73.49	45.93				
1966	67.60	94.43	45.93				
1967	58.40	95.90	45.93				
1968	51.10	96.64	45.93				
1969	51.80	101.78	45.93				
1970	56.60	103.62	45.93				
1971	58.10	107.66	45.93				
1972	81.90	110.97	45.93				
1973	165.70	124.56	45.93	114.94	112.80		
1974	154.30	75.32	50.34	127.93	125.10		
1975	137.40	75.32	50.34	139.44	136.45	76.02	74.39
1976	105.80	84.14	82.67	152.00	149.30		
1977	99.90	106.56	82.67	158.08	155.15		
1978	124.20	124.93	86.35	162.39	159.40	102.99	131.09
1979	156.20	124.93	91.86	201.42	197.45		
1980	163.50	133.38	110.23	214.01	209.20		
1981	156.90	139.99	117.58	230.55	225.55	142.63	139.54
1982	144.80	148.81	130.44	250.61	245.61		
1983	141.10	158.00	134.11	261.41	256.43		
1984	137.40	160.94	121.25	259.08	254.05	160.28	157.17
1985	120.50	160.94	121.25	254.98	249.95		
1986	99.90	160.94	88.18				
1987	108.80	160.94	83.78	255.10	251.39	168.04	164.95
1988		155.43	81.20	250.30	245.68		
1989		150.65	75.69	241.08	236.74	169.09	166.05

a No. 2 Hard Winter Wheat, Kansas City, ordinary protein
   Blanks indicate data not available
Sources: USDA, ERS, *Wheat Situation and Outlook Yearbook*, selected years; USDA, *ASCSCommodity Fact Sheet*, selected years; Commission of the European Communities, *The Agricultural Situation in the Community*, selected years.

The second point regarding the late 1970s and early 1980s is that as a result of these different policy responses, the U.S. lost market share (see

Table 8.3). The reason is very simple: to achieve its goals, the U.S. reduced production and exports, while the EC expanded in both areas in order to promote its objectives.

Although both regions were behaving in a desired fashion, the consequences for the U.S. were much different, and much more severe, than they were for the EC. Particularly troubling was that the U.S. found itself accommodating the EC. More specifically, as the U.S. reduced production and exports, thereby raising the world price, it became less expensive for the EC to raise its internal price, subsidize exports, and pursue its domestic policies. In addition, as the EC increased its exports and put a downward pressure on prices, the U.S. response was to further reduce its exports in an attempt to raise the price. The result was that the U.S. lost a substantial amount of the market share that it had obtained during the mid-1970s, and as a consequence, it changed its behavior.

While the maximization of producer returns less government expenditure was still a major U.S. goal, the method of achieving it changed abruptly with the 1985 Farm Bill. Fueled by the feeling that it was being taken advantage of by the EC and other exporting countries, the U.S. decided to adopt a policy that would make it expensive for its competitors to continue with their own policies. As discussed in the next section, the degree to which the U.S. is able to establish this new policy environment will determine whether or not an agreement is reached at GATT.

Figure 8.7 illustrates graphically the main features of the 1985 Farm Bill and their impact on the world wheat market. The most important elements of the bill were the lowering of the loan rate (in the case of wheat, it fell from \$U.S.137.40 per metric ton to \$U.S.120.50 per metric ton in 1985), the introduction of the Export Enhancement Program, the retention of the target price at historical levels (see Table 8.3), and the level of acreage set-aside.[9] With the retention of the target price at p_{us} ($= p_w^3$) and a continuation of acreage set-aside requirements, U.S. production continued at a level z_{us}^3. As a result of the Export Enhancement Program, however, the amount of output forthcoming from the U.S. effectively increased to z_{us}^4, which resulted in an increased level of exports (x_{us}^4) and a lower world price (p_w^4). In other words, the U.S. used its accumulated stockpile of wheat to increase its exports to x_{us}^4.

As a consequence of the lower world price, government expenditures in the U.S. increased by an amount $z_{us}^4(p_w^3 - p_w^4)$, i.e., the highlighted area in panel (a). Producer returns, however, remained relatively unchanged as a result of a continuation of the target price. The U.S. policy also had effects

Figure 8.7 The U.S. and EC Wheat Market: Mid-1980s versus the Late 1970s/Early 1980s

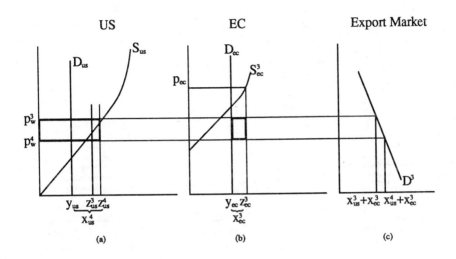

on the rest of the market. In the EC, producer returns remained unchanged, since the internal price (p_{ec}) did not change substantially. Government expenditures rose, however, as a result of the falling world price. In particular, EC expenditures on export subsidies rose by an amount equal to $x_{ec}^3(p_w^3 - p_w^4)$, i.e., the highlighted area in panel (b).

The area $x_{ec}^3(p_w^3 - p_w^4)$ can be seen as a measure of the cost that the EC has to incur in order to continue with its policy of setting the internal price at p_{ec}, while the area $z_{us}^4(p_w^3 - p_w^4)$ is the cost that the U.S. has to incur in order to try to influence the behavior of the EC. The magnitude of these two areas indicates that the U.S. has to incur a much greater cost than it can impose on the EC, implying that the U.S. may have to incur a substantial cost in order to get the EC to alter its behavior. It also suggests that the EC is unlikely to view the U.S. policy as credible, since the EC knows that it would always be worthwhile for the U.S. to revert to its original policy of stockholding. It is therefore unlikely that the EC would alter its behavior to any great extent in response to U.S. policy.

To show that it is committed to its policies and that its behavior is credible, the U.S. introduced legislation allocating Export Enhancement Program funds and requiring that they be used. In addition, by the late 1980s, the U.S. had reduced the acreage that farmers must set aside in order to be eligible for the target price. The results of this policy change are presented in Figure 8.8.

Figure 8.8 The U.S. and EC Wheat Market: Late 1980s versus the Mid-1980s

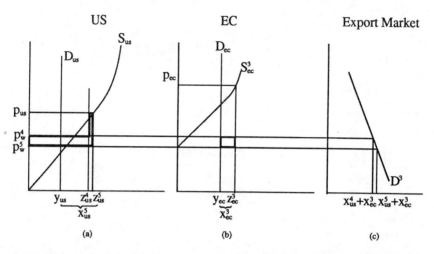

The price p_{us} represents the target price received by farmers. Relaxation of acreage set-aside requirements meant that U.S. farmers could grow as much as they desired at the target price, meaning, in effect, that the U.S. was abandoning its long-held position of reducing production in an effort to raise the world price. The result was that production increased from z_{us}^4 to z_{us}^5, exports increased to x_{us}^5, and the world price fell from p_w^4 to p_w^5.

As a result of this policy change, the returns received by farmers increased by an amount equal to the small triangle below p_{us}, to the left of z_{us}^4, and to the right of S_{us} (panel (a), Figure 8.8). At the same time, government expenditures increased by an amount $z_{us}^5(p_w^4 - p_w^5) + (z_{us}^5 - z_{us}^4)(p_{us} - p_w^4)$, and in the EC, they increased by an amount equal to $x_{ec}^3(p_w^4 - p_w^5)$. Both are shown by the highlighted areas in panels (a) and (b).

Prospects for a Trade Agreement

The discussion above illustrates that developments in the world wheat trade prior to the mid-1980s can be viewed as the outcome of a strategy by both the U.S. and the EC to maximize the returns to their farmers, less the government expenditures required to obtain those benefits. With the loss of

A New World Agricultural Order?

market share experienced by the U.S. in the early 1980s, the policy environment changed. In particular, by altering its behavior, the U.S. was attempting to initiate changes in the internal policies of other countries, specifically the EC.

The purpose of this section is to examine the prospects for a voluntary agreement between the U.S. and the EC to establish a new order in international agricultural trade.[10] Among the possibilities are (1) a move to free trade, and (2) a reduction in output, perhaps as a result of land conservation.

Figure 8.9 illustrates the impact of free trade on farmers and governments in the EC and the U.S., as compared to the situation in the late 1980s. Under free trade, the world price for wheat would be p_w^6. This price is determined by finding the intersection of the export demand curve (panel (c)) with the horizontal sum of ES_{ec} and ES_{us}, where ES_{ec} and ES_{us} are the excess supply curves for the EC and the U.S., respectively. The excess supply curves are found by horizontally subtracting the domestic demand curve from the domestic supply curve.

Given the demand and supply curves in Figure 8.9, the impact of free trade would be to raise the world price ($p_w^6 > p_w^5$). Since $p_w^5 > p_{us}$, farmers in the U.S. would be better off under free trade; the amount by which they benefit is given by the area between p_w^6 and p_{us} and to the left of S_{us}. In

Figure 8.9 The U.S. and EC Wheat Market: Free Trade versus the Late 1980s

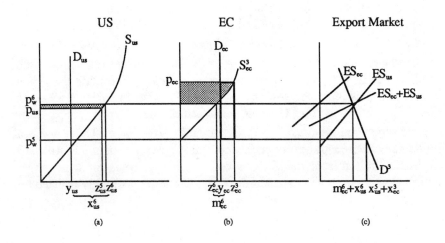

addition, the U.S. government would find its expenditures reduced by an amount equal to $z_{us}^5(p_{us} - p_w^5)$, thus benefiting in a substantial way from the introduction of free trade.

In the EC, however, the conclusion is not so straightforward. Farmers in the EC are unambiguous losers, since they receive the world price (p_w^6) instead of the higher internally set price (p_{ec}). The result is that they suffer a loss equal to the shaded area in panel (b). In contrast, the EC government benefits, since its expenditures fall by an amount equal to the highlighted rectangle in panel (b), i.e., $(z_{ec}^3 - y_{ec})(p_{ec} - p_w^5)$. As illustrated in Figure 8.9, it is unlikely that the gain by the government is sufficient to outweigh the loss by farmers. Thus, in terms of the goal of maximizing producer benefits less government expenditures, the EC is unlikely to want to adopt free trade. This conclusion is further strengthened when it is realized that the loss of benefits by producers would lead to severe financial hardships as land values fall throughout the EC.

This conclusion is dependent upon a number of factors, including the shape of the supply and demand curves, as well as the relative political power of producers, consumers, and taxpayers. With respect to the shape of the supply and demand curves, econometric estimates of the impact of a move to free trade indicate a wide range in the possible price increases — anywhere from 4 to 27 percent.[11] Some authors (see, for example, Furtan, Gray, Schmitz, and Ulrich) have argued that a reduction in prices in the EC would not influence output to any great extent, and that, as a result, a move toward free trade would not raise prices as much as it might otherwise. This would particularly be the case if the time period examined was short to medium term.

Figure 8.10 shows the impact of an inelastic supply in the EC on the prospects of the U.S. and the EC adopting free trade. Under free trade, the world price can be expected to be p_w^7; note that this is less than the free-trade price examined in Figure 8.9. In particular, p_w^7 is less than p_{us}, which implies that U.S. farmers would be worse off under free trade than they would be if they remained with the situation that exists in the late 1980s. In contrast, however, the U.S. government is better off, since their expenditures would have fallen by an amount $z_{us}^5(p_{us} - p_w^5)$. Thus, while a movement to free trade under the conditions described in Figure 8.10 increases the value of the U.S. objective function (maximum producer returns, less government expenditures), it is likely to be strongly resisted by farmers.

Figure 8.10 The U.S. and EC Wheat Market: Free Trade versus the Late 1980s, Inelastic Supply in the EC

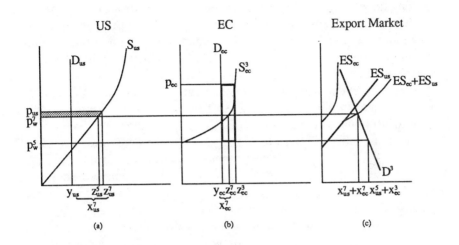

In the EC, a movement to free trade is unlikely to be supported by either the producers or the government. Farmers would find themselves substantially worse off; with the world price equal to p_w^7, farmers would lose an amount equal to the area above p_w^7, below p_{ec}, and to the left of S_{ec}. Although government expenditures would fall by an amount $(z_{ec}^3 - y_{ec})(p_{ec} - p_w^5)$, this would not be sufficient to offset the loss experienced by farmers. As a result, free trade would not be accepted by the EC in this situation.

The acceptability of free trade is also affected by the political power of consumers and taxpayers. In Figure 8.9, for instance, a movement to free trade and the subsequent lower price in the EC provides EC consumers with a benefit equal to $y_{ec}(p_{ec} - p_w^6)$. If consumers were to become a political force in the EC, this benefit might provide the EC government with some incentive to move toward freer trade.

Although consumers are unlikely to become a major political concern in the near future, events in Europe over the past year suggest that some change is likely to occur. More specifically, the end of the Cold War diminished the need for food security, while the political and economic restructuring that is occurring may highlight the need for less expensive food in eastern European countries.

In addition to consumers, taxpayers may also exert an influence on the decisions of the EC government. Historically, taxpayers seem to have been willing to support agricultural prices. This may have been partly because, although agricultural support spending has constituted 65-75 percent of the EC budget, it represents less than 1 percent of the EC GDP (Veeman and Veeman). With the reduction of political tensions in Europe and the increasing importance of other concerns such as the environment, taxpayers (like consumers) may begin to demand that agricultural programs provide other benefits besides food security. For instance, if these demands call for soil and water conservation, then the likelihood of agricultural policy changes increases. This will be examined in more detail below.

The conclusion that free trade is generally unacceptable to the EC (at least in the short run) implies that if the GATT negotiations are limited to this type of institutional structure, then there is little likelihood of a successful outcome to the talks. This, in turn, suggests that the departures from free trade that have occurred in the past should not be viewed as that unusual. More precisely, the above analysis suggests that when political and economic factors are considered, movements away from free trade are likely to be more desirable for a country or region than movements toward it.

The fact that free trade is not desirable for the EC does not mean that other political and economic orders are not possible. For example, Figure 8.11 illustrates a situation where both the U.S. and the EC reduce their production. Although there are many reasons why this may occur, it may be useful to view it as a response to the demand for increased environmental protection. In addition to providing soil conservation, output reduction

Figure 8.11 The U.S. and EC Wheat Market: Production Control versus the Late 1980s

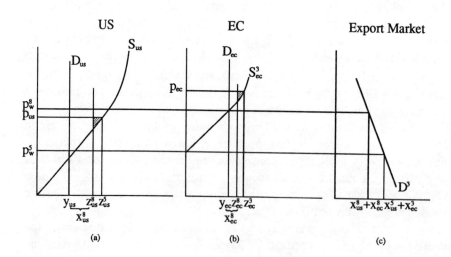

would contribute to less pollution (through reduced fertilizer and chemical use) and allow land to be put to alternative uses (e.g., recreation, green space).

As a result of the reduction in output to z_{ec}^8 and z_{us}^8, exports from the EC and the U.S. would fall to x_{ec}^8 and x_{us}^8, respectively. This would lead to an increase in price (p_w^5 to p_w^8). As a result, farmers in the U.S. would find themselves better off; their returns would increase by an amount $z_{us}^8(p_w^8 - p_{us})$, less the shaded area in panel (a). The U.S. government would also benefit, since exports would no longer have to be subsidized.

In the EC, farmers would be marginally worse off, losing the shaded area below p_{ec}, to the right of z_{ec}^8, and to the left of S_{ec}^3. The EC government, however, would be substantially better off. As a result of the increased world price, export subsidies would fall from a level ($z_{ec}^8 - y_{ec})(p_{ec} - p_w^5$) to a level ($z_{ec}^8 - y_{ec})(p_{ec} - p_w^8$). Since the decrease in government expenditures would be greater than the loss to farmers, there would be considerable pressure within the EC for the production cuts to be made. If farmers could be compensated for the reduction in output—through payments to take land out of production, for example, or to reduce fertilizer use—then the likelihood of production cuts being accepted would be greatly increased.

In the analysis undertaken above, the situation existing in the late 1980s is viewed as the relevant comparison point. The assumption is that the EC believes the U.S. would adopt the policies of this period if a new order could not be reached. However, if the EC believes that the U.S. behavior is not credible and cannot be sustained, then the comparison point must be changed. For example, if the situation in the late 1970s and early 1980s is viewed as the more likely alternative, then the EC should compare the benefits it receives under a new order with the benefits it had during that earlier period.

While the details of this comparison will not be presented, an examination of Figure 8.8 and Figure 8.5 shows that the main difference between the late 1970s/early 1980s and the late 1980s for the EC is an increase in government expenditures. The implication of this is that if the EC sees the late 1970s/early 1980s as the likely alternative to a new order, then it would be much less willing to accept a move to a new order, since expenditures could be reduced without having to move to free trade or restrict output. This suggests that the likelihood of a new order being adopted depends not only on the current situation, but on the degree to which this situation is credible and sustainable.

Conclusion

The first conclusion to draw from the above discussion is that the current situation—i.e., the current "order"—is a result of historical influences that embody both economic and political forces, and, in particular, the relative power of economic groups in society. The consequence of this is that any movement toward a new order has to encompass the issues of economic distribution. As a result, many potential orders have to be considered unreachable, since if the groups that have political power do not receive a benefit from a policy change, they will not accept it. As shown in the previous section, for instance, a movement to free trade is unlikely to succeed if the starting point is the current situation, where EC farmers have the major political influence and EC consumers are willing to pay a relatively high price for food.

The notion that some orders are presently impossible to attain does not imply that they will be forever unattainable. One of the conclusions to be drawn, in fact, is that the evolution of policy is an ongoing process. This suggests that the current situation, like all past situations, is in some way unstable. Indeed, the fact that agriculture is on the GATT agenda is an indication that countries are now willing to consider changes to the current order.

More specifically, while a new order in U.S. and EC agricultural trade is possible, a number of preconditions will have to be satisfied. Among these are a change in the willingness of EC consumers to pay the internal price—perhaps as a result of a change in the desire for food security—and the ability to solve farm income problems outside the market, e.g., through decoupled programs.

While political and economic forces will change—a good example is the recent shift in attitude toward the environment—it is important to recognize that the chances of reaching new agricultural orders will depend on the options being as varied as possible, suggesting that a wide variety of institutional arrangements will have to be considered.

The problems facing agriculture in the world today—whether it be the commodity surpluses in the developed world, the lack of food in the developing countries, or the inefficient use of resources in the centrally planned economies—suggest that the right institutions have not yet been found.

As an example, the discussion in the previous section highlights the confrontational nature of the current relationship between the EC and U.S. As long as they continue to behave in this fashion, the chances for an agreement will be reduced. The prospects for a new agricultural order that will meet the needs of all countries and regions will be enhanced as more cooperative arrangements are examined. A critical part of this will be the willingness of countries to recognize and place a value on the impact of their domestic policies on foreign countries.

The focus on the industrial countries in this chapter and the entire book points out another problem of developing a new world order in agriculture. In addition to recognizing the conflicts and interrelationships among the developed countries, the establishment of a new order must take into account the needs of the third world, as well as the environmental needs of all countries and regions. The new order will also have to cope with the current transformation of the centrally planned economies, which at present is a large uncertainty. In short, while the discussions in this book have focused on the industrialized countries, the questions that have been raised are even more important when seen in a global context.

Notes

1. The development of the wheat economy in western Canada is well covered in a number of sources, including Fowke and Wilson.
2. The one exception to this was the domestic wheat certificates (see the previous section).
3. Relaxation of this assumption would complicate the model without qualitatively changing the results.
4. In the discussion that follows, it is assumed that one price completely describes the EC's wheat price policy. The actual program is much more complicated than this. For a complete discussion of the EC's wheat price policy, including the importance and the impact of the threshold price and the intervention price, see de Gorter and Meilke (1987).
5. It should be pointed out that the U.S. was not the only country behaving in this manner in the 1960s. Canada also found it advantageous to restrict output during this period.
6. A more complete modeling of the supply shifts over time would have had the supply curve in both the U.S. and the EC shifting outward, with the EC shift being larger. To simplify the analysis, the differential is presented as a shift in only the EC supply curve.
7. See, for example, Penson and Fulton, Yeh et al., Spielmann and Weeks.
8. It should be noted that the gains in producer returns will depend upon the nature of the shift in the supply curve. For instance, an outward rotation of the supply curve would lead to a smaller increase in producer returns than the parallel shift presented in Figure 5.
9. Carter, McCalla, and Schmitz provide a good overview of the 1985 Farm Bill. For an analysis of the impact and effectiveness of the Export Enhancement Program, see Seitzinger and Paarlberg, and Paarlberg.

10. There are many elements to consider when examining the possibility of changes in agricultural policy. Runge and von Witzke discuss a number of the conditions required for institutional change and innovation in the EC's CAP, while de Gorter and Meilke (1989) analyze the relative efficiency of various policy options that have been proposed to deal with the problems facing the EC.

11. Carter, McCalla, and Schmitz provide an excellent overview of the various econometric models of the world grain trade that have been constructed, as well as summarizing the major qualitative and quantitative results.

References

Carter, Colin, Alex F. McCalla, and Andrew Schmitz. *Canada and International Grain Markets: Trends, Policies and Prospects.* Ottawa: Economic Council of Canada, Minister of Supply` and Services, 1989.

Commission of the European Communities. *The Agricultural Situation in the Community*, selected years.

de Gorter, Harry, and Karl D. Meilke. "The EEC's Wheat Price Policies and International Trade in Differentiated Products." *Amer. J. Agr. Econ.* 69.2 (1987), pp. 223-9.

———. "Efficiency of Alternative Policies for the EC's Common Agricultural Policy." *Amer. J. Agr. Econ.* 71.3 (1989), pp. 592-603.

de Hevesy, Paul. *World Wheat Planning and Economic Planning in General.* London: Oxford University Press, 1940.

Fennell, Rosemary. *The Common Agricultural Policy of the European Community.* Second edition. Oxford: BSP Professional Books, 1987.

Fowke, Vernon. *The National Policy and The Wheat Economy.* Toronto: University of Toronto Press, 1957.

Fulton, Murray, Ken Rosaasen, and Andrew Schmitz. *Canadian Agricultural Policy and Prairie Agriculture.* Ottawa: Economic Council of Canada, Minister of Supply and Services, 1989.

Furtan, H., R. Gray, A. Schmitz, and A. Ulrich. *International Wheat Markets: The Options Available to Saskatchewan.* Report submitted to the Government of Canada and the Government of Saskatchewan. Saskatoon: University of Saskatchewan, 1987.

Hadwiger, D.F. *Federal Wheat Commodity Programs.* Ames: The Iowa State University Press, 1970.

Harris, S., A. Swinbank, and G. Wilkinson. *The Food and Farm Policies of the European Community.* Chichester: John Wiley and Sons, 1983.

McCalla, Alex F. "A Duopoly Model of World Wheat Pricing." *J. Farm Econ.* 48 (1966), pp. 711-27.

Paarlberg, R.L. "The Mysterious Popularity of the EEP." *Choices.* (1990: Second Quarter), pp. 14-17.

Penson, John B., Jr., and Murray Fulton. "Impact of Localized Cutbacks in Agricultural Production on a State Economy." *W. J. Agr. Econ.* 5.2 (1980), pp. 107-22.

Polayni, K. *The Great Transformation.* New York: Octagon Books, 1975.

———. *The Livelihood of Man.* New York: Academic Press, 1977.

Runge, C. Ford, and Harald von Witzke. "Institutional Change in the Common Agricultural Policy of the European Community." *Amer. J. Agr. Econ.* 69.2 (1987), pp. 213-22.

Seitzinger, Ann Hillberg, and Philip L. Paarlberg. "A Simulation Model of the U.S. Export Enhancement Program for Wheat." *Amer. J. Agr. Econ.* 72.1 (1990), pp. 95-103.

Smith, Adam. *An Inquiry Into the Nation and Causes of the Wealth of Nations.* Homewood, Illinois: Richard D. Irwin, Inc., 1963.

Spielmann, Heinz, and Eldon E. Weeks. "Inventory and Critique of Estimates of U.S. Agricultural Capacity." *Amer. J. Agr. Econ.* 57 (1975), pp. 921-28.

Tracy, Michael. *Agriculture in Western Europe, Challenge and Response 1880-1980.* Second edition. London: Granada, 1984.

United States Department of Agriculture, Economic Research Service. *Wheat Situation and Outlook Yearbook*, selected years.

———. *ASCS Commodity Fact Sheet*, selected years.

———, Foreign Agricultural Service. *World Grain Situation and Outlook*, selected years.

Wilson, C.F. *A Century of Canadian Grain.* Saskatoon: Western Producer Prairie Books, 1978.

Yeh, Chung J., Luther Tweeten, and Leroy Quance. "U.S. Agricultural Production Capacity." *Amer. J. Agr. Econ.* 59 (1977), pp. 37-48.

SECTION FOUR

Conclusion

9

Concluding Remarks

Hans J. Michelmann and Jack C. Stabler

The Importance of Political Factors in International Agricultural Trade

When the editors planned the conference at which the papers constituting this volume were presented, they deliberately asked contributors to emphasize an element that has often received insufficient attention in discussions of agricultural trade. This was the *political* considerations, which are of major importance alongside or in combination with the economic factors that have been the traditional focus of such symposia; hence the title of both conference and book: *The Political Economy of North American-European Agricultural Policy and Trade*.

The utility of emphasizing the political as well as the economic aspects of agricultural policy and trade has been amply demonstrated in the preceding chapters. Indeed, political factors are of major importance for understanding a number of items: a) important domestic determinants of the positions taken by the participants in international trade negotiations, since these participants are, if not politicians themselves, working under instructions from elected politicians either directly, as in the U.S. and Canada, or indirectly, as in the EC; b) the nature of the way in which decisions are made at the national level and, for the EC, in the Community context; and c) the nature of decision-making in the international arena (the GATT), where new trade policies are hammered out in the interactions among the representatives of participating states and trading blocs.

The domestic determinants of agricultural decision-making are among the most crucial variables that must be taken into account in order to make sense of the European Community's Common Agricultural Policy (CAP) and its apparent irrationality in economic terms. As Michael Tracy points out in his chapter, a major factor in the process that led to the founding of the EEC in

1958 was the political bargain between the two giants among the original member states, France and the Federal Republic of Germany (F.R.G.). This bargain was to lead to unimpeded intra-Community trade in industrial commodities, from which the F.R.G. would benefit at the expense of what the French considered their vulnerable factories. In return, the common agricultural policy would disproportionately benefit France by providing her comparatively efficient farmers with a large and protected export market for their produce. When the Community opted for a price support system of assuring farmers' incomes, when these prices were set at levels sufficiently high to also allow less efficient German farmers a reasonable return, and when these measures were coupled with border controls to protect EC farmers from lower-priced agricultural commodities on the international market, a policy was established that became France's "ark of the covenant" in the Community and one provided farmers elsewhere in the original northern member states a similar vested interest.

The costs of this policy were borne within the EC by less well-organized interests, viz., the undifferentiated mass of consumers/taxpayers. Costs of the CAP also impinged on outsiders, especially when production of certain commodities, spurred on by high Community target prices, began to soar. At that point, third-country farmers with no votes in EC member states not only lost traditional European markets, but saw commodity prices on the international market fall in response to EC domestic oversupply, which was then reduced somewhat by subsidized exports.

As Michael Tracy points out, the fact that this system of subsidization has not been seriously questioned in the EC until recently, attests not only to the durability of the national interests of those member states benefiting from the CAP, and to the fact that the CAP costs only a small fraction of Community GNP, but also to the key role of farming constituencies in the political calculations of governments. These interests sometimes can induce governments to take positions that are not to the advantage of the country as a whole. The most striking example is the F.R.G., a member state that is a large net importer of agricultural commodities, where the general welfare (though clearly not that of the small agricultural sector) would clearly be served by low farm prices. The Bonn government, however, has supported a high price policy in the annual EC agricultural ministers' negotiations because of the alliance of the German farm lobby with key elements in the F.R.G.'s governing coalition. With farm prices a nonissue in national politics, and its future political prospects in the balance because of wafer-thin majorities in parliament, it is to the government's advantage to benefit this narrow constituency at the expense of the great mass of voters. The wider implication of such behavior is to make CAP reform more difficult, but it is necessary if international agricultural trade agreements are to lead to a more rational trading regime. The paradoxical German behavior on CAP prices is only understandable when these domestic political factors are taken into account.

Changing EC agricultural policy has been difficult in the past not only because of the strength of vested interests, but also because the EC decision-making system tended towards the consensualism characteristic of international organizations, rather than the majority-rule style in effect for national governments. Hence, though in recent years member state governments have allowed themselves to be overruled in routine agricultural negotiations, on matters of vital national interest to one or more of them, majority rule does not apply automatically. And the concept of "vital national interest" is sufficiently ill-defined that a determined member state, particularly one of the larger ones, can scuttle an agreement if it will have disagreeable domestic consequences. This means, in turn, that the EC's position on international agricultural trade issues can be held to ransom by one or two member states. This political reality must give pause to anyone hoping for wide-ranging CAP reforms designed to expedite agreement on a new international agricultural trading regime, if these reforms have a strong negative impact on the farming community in one or more EC settings.

It must also not be forgotten that the CAP has a social policy dimension. The policy of maintaining comparatively high domestic prices must be continued if the large number of small, marginal farms in the EC are to keep even a semblance of viability. As in North America, the concept of viability is infused with and motivated by the desirability of maintaining the family farm; but more than in North America, where the average farm is a reasonably efficient operation, European rationalization implies economic dislocation of a large number of marginal operators. This is so, for example, in parts of southern Italy, and in Portugal, Spain, and Greece, four countries struggling to modernize their economies and deal with high unemployment rates at the same time. Even in the developed north-central industrial core region of the EC, unemployment rates are high and governments are reluctant to displace marginal farmers, who would have great difficulty finding places in an economy that is increasingly sophisticated and thus requires specialized skills not found among the rural population. In a political climate already destabilized in some countries by extremist parties such as the Republicans in Germany and the National Front in France (who make their appeals to the society's marginal elements), increasing rural to urban migration is not a pleasant prospect for political leaders.

There are yet other political factors at work. Grace Skogstad, as well as Gary Storey and Murray Fulton, points out that the question of security of supply must be taken into account when considering the positions different countries will take on agricultural questions. Europe has on at least two occasions in this century faced the prospect of, or even experienced, widespread undernourishment, if not starvation, when food imports were interrupted by war. Such considerations led to one of the explicitly stated goals of the CAP, namely, secure food supplies, and have contributed to the thrust towards higher production and its trade-diverting effects. It is not easy for North Americans, who have not in living memory collectively faced major food shortages, to grasp the importance of these motivations.

One final political factor relevant to the EC must be mentioned here, and not only because of its dramatic currency. That is the political and economic ferment in Eastern Europe and the Soviet Union. Its impact on decision-making in the EC is most immediate in the F.R.G.'s present preoccupation with reunification and its concomitant decreased attention to the EC. As Michael Tracy argues, in this climate West Germany's EC partners are unlikely to want to initiate policies (e.g., dramatic price reductions in agriculture) that will destabilize that country's domestic politics. Further, it is still unclear what impact EC agricultural purchases from Eastern Europe, motivated by the desire to help these countries weather the turbulence of political and economic reform, will have on international agricultural trade. It is also difficult to predict the impact of EC food aid to these countries while they are struggling to reorganize their agricultural sectors along private lines, because one cannot be certain what dislocations lie ahead, or how much time the process of adjustment to new conditions will take. Finally, in the medium to long term, one must also take into account the increased agricultural production in Eastern Europe and the U.S.S.R. that will result from the greater efficiency of restructured economies and altered incentive structures. Will this result in increased agricultural surpluses, and if so, of what commodities? Or will the increase in living standards that is a likely result of economic reform lead to increased demand for high quality foodstuffs and thus to increased imports? All these factors have implications for the larger agricultural trade context.

Political factors play an equally significant role closer to home, in North America. Grace Skogstad neatly summarizes the varying policy configurations in Canadian agriculture, each of which is characterized by its own constellation of interest-group politics, both with respect to the interaction among the agricultural groups, and in terms of the dynamics of their relationships with government. Her discussion demonstrates the complexities of agricultural policy-making in Canada. Canada's federal government faces three differently functioning policy sectors; provincial governments, with their own jurisdictions and interests; a pluralistic, fractured, and often conflict-ridden interest-group structure without an effective central leadership; and a lack of consensus among important actors even within certain sectors on issues salient to international negotiations. As a result, Canada's federal government may well have more leverage now to take action domestically with respect to any one of these actors, than previously, when the interest-group structure was less fractured, but arriving at a coherent policy on international agricultural trade is no easier, and probably more difficult.

Gordon Rausser also highlights the importance of political factors for understanding U.S. agriculture. What is striking in his work is the strength of the theoretical and historical arguments that highlight the intricacies of the processes whereby economically defensible PERT policies were diverted, one may say perverted, into PEST policies by the political shrewdness of self-serving agricultural groups, agro-industries, and politicians. The complexities of American governing institutions, the separation of powers, with its loose or practically absent party discipline, and the multiple entry points for lobbying that such a system provides, makes prognoses regarding policy outcomes in the U.S. almost impossible. It is difficult enough in parliamentary systems such as the Canadian, in which a decisive executive will get its way in the legislature.

Yet other features of domestic politics introduce additional complications to agricultural policy-making. Environmental issues and "green" political movements or even political parties, significant parts of the political landscape for years in some European countries such as the F.R.G., are becoming increasingly relevant in North America as the chapters by Skogstad and Runge demonstrate. The dynamics of the development of green concerns and political movements, with the attendant emphases on the national political agenda of environmental and health concerns, are not well understood, nor are their potential impacts on policies, but they have caught the attention of politicians and caused them to compete with each other to be "greener than thou." In Western Canada and elsewhere, voices are being raised against present chemically-intensive farming practices, yet lesser reliance on chemicals will have implications for production. Further, the regulation of pesticides and other chemicals may well give advantages to exporters of farm commodities in countries with less stringent environmental regulations, leading elsewhere to calls for protection by farmers subject to what they consider to be unfair competition from abroad. If, as Ford Runge convincingly argues in his chapter, environmentally-motivated, nontariff barriers become more prominent in the future, political factors will become even more important in agricultural trade negotiations, and these will become even more complex than they already are.

Tim Josling's chapter demonstrates the difficulties that have faced such negotiations in the past, and the problems of achieving a comprehensive agricultural trading agreement in the present Uruguay Round. The fundamental issue bedevilling these negotiations is the classic one facing international organizations and international relations generally: the unwillingness of participating states to allow international actors and processes to interfere in domestic matters. The resulting hiatus between what will provide a generally acceptable resolution in the best interests of the world community as a whole, and the adjustments within states required to bring about such a solution, is difficult to bridge in an international system organized around sovereign states in which politicians rely for their tenure on domestic constituencies. The existence of national, and for the EC, regional, governmental structures between the international institutional arena, such as that inhabited by GATT, and the application of domestic farm policies is the major reason that, in the words of Professor Josling, "it is not easy to find examples of domestic policy changes stemming directly from GATT rulings." His chapter provides evidence that there are no angels among the three major actors discussed in our volume: Canada, the EC, and the U.S. have all undertaken actions that, while understandable in domestic political terms, have negative implications for international agricultural trade.

In the light of historical experience, there appears little reason to hope that a solution will be found to what Gordon Rausser has identified as the prisoner's dilemma game, in which the players would be collectively better off by cooperating, but individual players stand to lose drastically if they adopt cooperative strategies while their fellow players choose not to. Recognizing this dilemma and also the fact that it is not likely that governments abroad will cease subsidizing farmers, Premier Devine of Saskatchewan has been prompted to argue that Canadian farmers, particularly those in Saskatchewan, must be supported so that they can survive the present subsidy war and not leave international agricultural trade by default to farmers who, in the end, produce less efficiently than their Canadian counterparts. In economic terms and in terms of broader conceptions of justice, bankruptcy of Canadian farmers resulting from subsidization of less efficient producers elsewhere is not a desirable solution.

One additional factor should be taken into account when considering the prospects for the attainment of an international agreement to structure agricultural trade more rationally. Professor Rausser points out that the

U.S. Congress is likely to consider all matters under negotiation in the present GATT round as a package, and not to consider agricultural issues separately. It is difficult to predict whether the binding of these separate issues into one over-arching decision, a feature of American politics, will make the attainment of an agricultural settlement easier. It probably will not, given the current climate of U.S. aggrievement with its international trading partners on many fronts, not just agriculture.

This summary, and by no means comprehensive, discussion of some political factors impinging on agricultural trade demonstrates their centrality to the matters at hand, and the complexity they add to negotiations aimed at bringing about greater rationality in that trade sector. It is difficult not to adopt the "realist" scenario outlined by Tim Josling in chapter 7, and hence not to be pessimistic about the outcome of the agricultural trade talks in the present GATT round, because there is comparatively little evidence that the parties with a vested interest in the present trading regime can be induced to agree to change. For Canadians this is a disconcerting conclusion, not only because agricultural products comprise a substantial proportion of our exports, but also because in the international arena, given Canada's limited clout, there is not a great deal the country can do to affect the outcome. It is not even clear with which states Canada should ally herself to bring about an outcome favorable to her interests. For Saskatchewan, such a conclusion is even more gloomy because of the central role that agricultural trade plays in the provincial economy. Jack Stabler's remarks in the following section focus on the implications for Saskatchewan of a positive outcome of the GATT trade talks, and also on what should be expected if these talks fail.

The Economic Dimension: Saskatchewan and the Outcome of the GATT Negotiations

Examining the implications of the outcome of the GATT process for Saskatchewan is an exercise in forecasting. Anyone who engages in forecasting attempts to develop a perception of the future on the basis of observable, systematic influences, and in this regard, there are some important trends. It is known, for example, that the western world, and more recently, some countries in the third world, can produce more grain products than the market would like to take away at prices that would compensate farmers for producing it. It is also known that technological change, which has led to increased production in the third world, and in western economies as well, will continue to make contributions to crop yields over the next several years. Looking, then, at the growth of effective demand, the consensus appears to be that our ability to produce will exceed the rate at which demand is forecast to grow over the next several years—perhaps over the next decade. Building a systematic forecast around these trends leads to certain inevitable conclusions. There are, of course, unsystematic influences that can affect this forecast, drought being probably the most important one over the long haul for grain-producing areas. In addition, there is now a new possibility to include in this equation. This is the political change in Eastern Europe. It would not be surprising if this political change negatively affected the ability of Eastern Europe or the USSR to produce, at least over a short term, possibly a two-to-four-year period. In a long-term context it is necessary to contemplate what Eastern Europe and the USSR are likely to do with their productive capacities, probably in terms of decades, or a decade at least, rather than years.

So with that as a basis for forecasting, what can be expected out of GATT? Dwelling on the systematic influences, and assuming that unsystematic influences are not going to have an impact over the next several years, the following implications seem warranted. It seems most likely that for Saskatchewan, whether GATT is successful or not, will not make much difference in the short run, though it clearly will in the long run. Let us first deal with the implications of a successful conclusion. Professor Storey's chapter provides the background for the evolution of trade policy in Western Europe and the U.S. It took the best part of a century to move from generally unrestricted trade to the pervasive pattern of supports and protection in place now, implying that the structure has evolved over decades of increasingly greater protection. A quick fix is unlikely because of the massive disruptions and restructuring this would cause. Further, both the U.S. and Europe are wealthy enough to continue the subsidies. They may be large in absolute terms, but in comparison to the wealth of those economies, they are not great. In neither area, as Michael Tracy and Ford Runge pointed out, is there overwhelming public demand for change. Farmers in particular do not want changes, or at least not rapid changes, the general public seems ambivalent, and no pressure group is agitating for reform. Still, the major participants in the trade war, and those affected by it, could use the billions of dollars going into export enhancement, or PEST-type programs, in alternate ways.

There does seem to be a basis for cautious optimism, nevertheless, that some kind of agreement may come out of the GATT process. What are we to anticipate? Probably the best that can be hoped for is an agreement to put a lid on the support programs, as the Saskatchewan premier suggested recently, and a consensus on a framework within which, over a fairly long

Concluding Remarks

time, the market will eventually have an ever-increasing influence. This would, in fact, be a substantial accomplishment.

What does "success," defined in this way, imply for Saskatchewan? The most positive interpretation, given success, is that there would appear to be a light at the end of the tunnel. But it is a very long tunnel, as long as ten years perhaps. Even that may be optimistic; it could be longer. This view differs from that held by some residents of rural Saskatchewan, where there are people who think that success will lead to a rapid increase in prices. Success, to them, means that in three or four years, prices would edge back up to five dollars a bushel. It should be fairly clear, however, that this is not in the cards. During the time it takes to pass through the tunnel, given that unsystematic influences do not lead to a reduction in supply, wheat prices, in real terms, will likely be well below this $5 level.

How, then, does Saskatchewan respond? The province must adjust to the long-term prospect that wheat prices in the future will not be greatly different from those of the present. This has numerous implications—for farm size, input mix, debt load, cropping practices, and so forth. Recognition of this implies that we need to move away from solutions developed on a year-by-year basis. This approach was unavoidable, of course, when it was not clear how long the process would continue, but there is now a much better perspective on that. We must continue to look for ways to diversify Saskatchewan's agricultural product mix within the context of our location and our resource base, both of which impose numerous constraints. We must also find ways to diversify other sectors of the provincial economy as well. We need to create nonfarm jobs in rural areas, i.e., in-town jobs in rural towns; jobs for farmers, particularly those who could perhaps make a go of it if they had an additional source of income to tide them over the first few years; and nonfarm employment for young farmers trying to get established. In addition, we need to develop public policies that will ensure that we retain a network of viable rural communities, because these communities serve the farmer both as producer and consumer. They will also be the places in which any nonagricultural employment is generated in rural Saskatchewan.

We probably will need training and relocation programs for those forced out of agriculture. One third of Saskatchewan farmers are in severe financial difficulty at the moment; many of them will not complete the journey through the tunnel. In addition, other rural dwellers, such as owners of small businesses, will also be forced into bankruptcy as agriculture and the rural trade system contracts. We have 598 incorporated communities in Saskatchewan, only 60 of which are clearly viable; over 500 are not. This lack of viability ranges from moderate to terminal. It is clear that many of the people who own businesses in those communities will be forced out. They, as well as many farmers, may require retraining and relocation support.

Our federal government's role in all of this will be to establish the framework within which Saskatchewan's adjustment occurs, a policy process that is under review already. The evolving policies, as they apply to grains, recognize the situation for what it is, and an attempt is being made to define a path of adjustment to long-term realities. But the federal government should not put away the cheque book yet. Saskatchewan agriculture, indeed the province's entire economy, is not viable in the context of today's wheat prices. The grain economy will need an injection of several hundred million dollars a year for the next several years. Even a half-billion dollar addition to Saskatchewan's economy would not make it viable. It would just provide the funding necessary to facilitate the kind of adjustment we have to consider. Through time, as the economy restructures, the level of support can diminish and can be increasingly decoupled.

What if no agreement is possible during the present round of GATT negotiations? Subsidies of all sorts will continue, grain prices will remain low, Saskatchewan agriculture and the entire provincial economy will continue to be under stress. The difference, in the short run, of a successful and an unsuccessful outcome at GATT is that with a successful outcome there is a light at the end of the tunnel; with an unsuccessful outcome there is only the tunnel.

Can we conclude on a positive note? There does seem to be sufficient overlap in the present and anticipated U.S. and EC positions to generate a degree of optimism. We have collectively concluded at this conference that it is unpleasantly expensive and wasteful to continue down the present path. We can hope, perhaps even anticipate, that these considerations will produce a framework within which to proceed to dismantle existing trade barriers. If we in Saskatchewan know what these rules are, that is, if there is a framework for agreement, and if we have an indication of the time frame within which certain objectives will be realized, we can begin to adjust to an environment in which there is greater stability than what has characterized the past several years. The Saskatchewan economy is resilient enough to successfully pursue such an adjustment, even in the context in which grain prices are lower for the first several years than what would prevail in a completely market-determined environment, given, of course, that there is continuous federal financial support during this adjustment process.